ONE
HUNDRED
PROBLEMS IN
ELEMENTARY
MATHEMATICS

ONE HUNDRED PROBLEMS IN ELEMENTARY MATHEMATICS

HUGO STEINHAUS

With a Foreword by
Martin Gardner

Dover Publications, Inc.
New York

Published in Canada by General Publishing Com-
pany, Ltd., 30 Lesmill Road, Don Mills, Toronto,
Ontario.

This Dover edition, first published in 1979, is an
unabridged and unaltered republication of the
English translation first published in 1964 by Basic
Books, Inc. The Dover edition is published by
special arrangement with Basic Books, Inc., 10 E.
53rd St., New York, N.Y. 10022.

International Standard Book Number: 0-486-23875-X
Library of Congress Catalog Card Number: 79-52010

Manufactured in the United States of America
Dover Publications, Inc.
180 Varick Street
New York, N.Y. 10014

CONTENTS

Foreword
Preface

Chapter V

PROBLEMS ON CHESS, VOLLEYBALL AND PURSUIT

Chapter VI

MATHEMATICAL ADVENTURES OF DR. ABRACADABRUS

Chapter VII

PROBLEMS WITHOUT SOLUTION

FOREWORD

MARTIN GARDNER

"The sheep and the goat", said Abraham Lincoln, "are not agreed upon a definition of the word liberty". One suspects that the professional mathematician and the layman have a similar mismeeting of minds over the word "elementary". Are the problems in this collection "elementary", as the book's title indicates? They are in the sense that the word is used in the section on "Elementary Problems" in *The American Mathematical Monthly,* that is, problems that do not call for a knowledge of calculus or higher levels of mathematics. But, if the average layman reading this book expects to *solve* most of its problems, he must know thoroughly his mathematics up to calculus. Above all, he must be able to think clearly and creatively.

Does this mean that readers of lesser skill and knowledge cannot read the book without enjoyment and profit? Far from it! Professor Steinhaus has given complete, detailed answers to every one of his one hundred problems. A layman who does nothing more than turn immediately to the solutions and then think his way through them will find himself learning painlessly, almost without realizing it, an astonishing amount of significant mathematics. The best way to use this book, if you find most of the problems too difficult, is to take the difficult ones slowly, perhaps no more than one or two a day. If you make no headway with a problem, turn to the answer and read it carefully, several times if necessary, until you fully understand it. Do not skip the problem just because you encounter unfamiliar terms or procedures. Try to find out from other books or from friends what the terms and procedures are all about.

The literature of recreational mathematics is enormously repetitious. A great merit of Professor Steinhaus' collection is that the problems are not only top-drawer, but that most of them are relatively new, not to be found elsewhere. I know of no book of solved problems on this "elementary" level so refreshingly free of the old chestnuts that usually abound in collections of this type.

Another great merit of this book is that its problems often have elegant, sometimes totally unexpected, solutions. What a delight to discover, for instance, that Problem 97, which seems to deal only with relations between friends in a social set, actually involves the structure of the dodecahedron! And how beautifully clear certain confusing problems become after the author has drawn a relevant graph. I think particularly of the graph for that bewildering problem of the watch with two identical hands (Number 58), the diagram of the cyclist's minimum path for catching two pedestrians going in opposite directions (Number 86), and the startlingly simple graph that proves the constant length of a ribbon tied around a box in a certain way (Number 66). Many department stores, by the way, now fasten boxes with an endless loop of slightly elastic tape in just the manner given in this problem. Such a method is efficient for precisely the reason disclosed in the problem's answer.

The more skillful reader, if he enjoys puzzle-solving, will find that many problems in this book can be generalized in interesting ways. Consider Number 87, in which four dogs, starting at the corners of a square, chase one another simultaneously. What happens if the dogs start at the corners of an equilateral triangle? Of a pentagon? Can you find a general formula for any regular polygon? Similar questions arise in connection with Problem 73. After you have mastered the proof for the minimum network joining the four corners of a square, see if you can find the minimum network for the five corners of a pentagon. A very pretty problem in three-space is suggested here, and one I have not yet seen in print. What is the minimum network joining the eight corners of a cube?

Problem 33 provides a beautiful solution to a problem once proposed by Lewis Carroll. Warren Weaver, in his article on "The Mathematical Manuscripts of Lewis Carroll," *Proceedings of the American Philosophical Society*, 98 (1954), No. 5, expresses it this way: "Can a billiard ball travel inside a cube in such a way that it touches all faces, continue forever on the same path, and all portions of the path be equal?" Steinhaus does not add the last proviso, but, if the ball's chair-like path is adjusted so that it strikes each side at a point of ⅓ distance from one edge and ⅓ from the other edge, the proviso is met. Each segment of the path has a length of $1/\sqrt{3}$. This was worked out in 1960 by Roger Hayward, who illustrates Red Stong's "Amateur Scientist" department in *Scientific American*. (See Hayward's article on "The Bouncing Billiard Ball" in *Recreational Mathematics Magazine*, June 1962).

Several years ago, Professor Steinhaus proposed the similar problem of finding an equal-segment solution for a ball bouncing inside a regular tetrahedron. This also was solved by Hayward, early in 1963, and published in the "Mathematical Games" department of *Scientific American* in September, 1963. Are there such paths inside the other regular polyhedra?

Problem 34 (one of the few of the one hundred that is not new; Henry Dudeney gave it as Problem 146 in *Modern Puzzles* [1926]) suggests an interesting extension into four dimensions. This occurred to me one day when I stopped at a store window in New York to look at a color reproduction of Salvador Dali's remarkable painting of the crucifixion, *Corpus Hypercubus*. (The original is owned by the Metropolitan Museum of Art, New York; reproductions are obtainable from the New York Graphic Society.) The figure of Christ hangs on a three-dimensional cross formed by eight cubes. Dali clearly had in mind an "unfolded" hypercube. The three-dimensional cross floats above a flat, two-dimensional checkerboard. Light from an infinite source casts a shadow of Christ on the unfolded hypercube, suggesting that the historical event is a projection of a transcendent event onto the space–time continuum of our world. Just as the cube can be unfolded in many ways to make a flat figure of six squares joined by their edges (a "hexomino"), so can a hypercube be unfolded in many ways to make a solid figure of eight cubes joined by faces. A cube has six square faces joined at twelve edges. Seven of these edges must be cut to unfold the squares. The hypercube has eight cubical hyperfaces joined at twenty-four faces. To "unfold" it into three-space, seventeen of these faces must be cut, leaving the cubes joined at seven faces.

In how many ways can this be done? The problem should not be difficult for mathematicians who know their combinatorial way around in four-space.

Pictures of and additional details on some of Professor Steinhaus' problems can be found in the revised edition of his wonderful book, *Mathematical Snapshots*, (Oxford University Press, 1960). I can recall the delight, some twenty years ago, with which I examined the first edition of this book, printed in Poland in 1938. It was crammed with paper and cardboard gimmicks. Graphs drawn on transparent paper were tipped-in over pictures. An envelope at the back of the book contained such things as red and blue spectacles for three-dimensional viewing of certain illustrations; a cardboard model of a dodecahedron threaded with elastic to make it pop into solid shape; a cardboard torus colored with seven colors, each bordering the other six; a packet of thirty-one cards that could be assembled in a certain order, then flipped at either end, on both sides, to make motion pictures of a bullet travelling a parabolic path, a planet moving in an ellipse, a circle rolling inside a larger circle to generate a hypocycloid, and a ball racing down the curve of quickest descent ahead of another ball on an inclined plane.

Hugo Dynoizy Steinhaus was born in 1887 at Jaslo, Poland, and trained in mathematics at Göttingen University, where he received his doctorate. At present he is professor emeritus at the University of Wroclaw (Breslau) and one of Poland's most distinguished mathematicians. He has published some 150 papers on pure and applied mathematics, edited mathematical journals, and received many mathematical awards. His interest in recreational mathematics is lifelong and unbounded. In the preface to the first edition of *Mathematical Snapshots,* he stated that the book's gimmicks and haphazard arrangements were designed to appeal "to the scientist in the child and the child in the scientist". "Perhaps", he concluded, "I have succeeded only in amusing myself".

The same spirit of play pervades this little book of problems. Dr. Steinhaus can rest assured that *Sto Zadan* (as the book is called in Poland) will, like his previous book, both amuse and educate many thousands of kindred souls in the English-speaking world.

●

A note on notation. This British translation from the Polish has retained the metric system of measurements used in the original. Abbreviations for metric units should cause the reader little difficulty: mm = millimeter, km = kilometer, cm = centimeter, km² = square kilometers, and so on. The decimal point in British books is traditionally raised from the base line. This is confusing to United States readers, especially because the same raised point is sometimes (as in Problem 1) a symbol of multiplication.

PREFACE

This booklet is an answer to a challenge: a few years after the war the inadequacy of mathematical education in our high schools became evident to the staffs of universities and technological institutes. Some responsible people felt that a closer collaboration between mathematicians and school teachers could no longer be postponed. A few scientists were among those who did their best to stimulate interest in mathematics by means of elementary problems published in an educational journal. Here the reader will find one hundred elementary problems and their solutions. Some of them are familiar to students in high schools, but it was by no means my intention to provide the teacher with questions he could find in every textbook. I have tried rather to formulate problems suggested by geometry, often without classifying their mathematical background. As higher mathematics is not supposed to lie within the reach of the reader I was limited in my choice and this explains the small size of this collection. The solutions, however, are detailed enough to be understood by the teacher and by those of his pupils who are not afraid of thinking. Some of the solutions have been found by readers of the bimonthly journal referred to above — their names are printed on page 196 of the Polish edition. The last chapter has a few questions without corresponding solutions. There is an extremely important reason for such an omission for at least some of the thirteen items of Chapter VII: The author does not know the solutions; he hopes that his readers will try to solve some of them, thinking that their solutions are known, and that this mistaken view will enable them to succeed where the author has failed.

The "Hundred Problems" may help some freshmen discouraged by the difficulties of higher mathematics. Showing them elementary mathematics from another point of view, the "Hundred Problems" tries to bridge the apparent gap between "elementary" and "higher" mathematics. The book appeared first in Polish; I have been helped essentially by Dr. S. Paszkowski in the preparation of this first edition. The English version now presented to the reader is a revised edition. The translation into English has been accomplished by Mr. Bharucha-Reid in collaboration with Miss R. Czaplińska, Mrs. J. Smólska and Mr. H. Brown. The author is very much obliged to all the persons named above, especially to Mrs. Smólska, who spared no effort to make this little book readable by the English-speaking public. There is also a Russian translation, of which 100000 copies have been printed.

HUGO STEINHAUS

Wrocław

ONE
HUNDRED
PROBLEMS IN
ELEMENTARY
MATHEMATICS

CHAPTER I

PROBLEMS ON NUMBERS, EQUATIONS AND INEQUALITIES

1. Exercise on the multiplication table

We construct a sequence of numbers as follows: The first number is 2, the next is 3,

$$2.3 = 6,$$

the third number of the sequence is 6,

$$3.6 = 18,$$

the fourth number is 1, and the fifth is 8,

$$6.1 = 6, \quad 1.8 = 8,$$

the sixth number is 6, then follows 8, etc.

This is the sequence which we obtain:

$$2\smile3\smile6\smile1\smile8\smile6\smile8 \ldots$$

The little arcs under the numbers denote the multiplication carried out, the result of which follows the last digit of the sequence. For example, we ought to multiply now 6 by 8 and write down the numbers of the result, namely 4, 8. There will never be a shortage of numbers for multiplication, since the number of arcs is increased by one with each multiplication and the result will yield at least one and often two digits, so that there always appears at least one new digit.

Prove that numbers 5, 7 and 9 never appear in this sequence.

2. An interesting property of numbers

Let us first write an arbitrary natural number (for example, 2583), and then add the squares of its digits $(2^2+5^2+8^2+3^2 = 102)$. Next, we do the same with the number obtained $(1^2+0^2+2^2 = 5)$, and proceed in the same way $(5^2 = 25, 2^2+5^2 = 29, 2^2+9^2 = 85, \ldots)$.

Prove that unless this procedure leads to the number 1 (in which case the number 1 will of course recur indefinitely, it must lead to the number 145, and the following cycle will occur again and again:

$$145, \ 42, \ 20, \ 4, \ 16, \ 37, \ 58, \ 89.$$

3. Division by 11

Prove that the number

$$5^{5k+1} + 4^{5k+2} + 3^{5k}$$

is divisible by 11 for every natural k.

4. The divisibility of numbers

The number

$$3^{105} + 4^{105}$$

is divisible by 13, 49, 181 and 379, and is not divisible either by 5 or by 11.

How can this result be confirmed?

5. A simplified form of Fermat's theorem

If x, y, z and n are natural numbers, and $n \geqq z$, then the relation $x^n + y^n = z^n$ does not hold.

6. Distribution of numbers

Find ten numbers $x_1, x_2, \ldots, x_{10}$ such that

(i) the number x_1 is contained in the closed interval $\langle 0, 1 \rangle$,

(ii) the numbers x_1 and x_2 lie in different halves of the interval $\langle 0, 1 \rangle$,

(iii) the numbers x_1, x_2 and x_3 lie each in different thirds of the interval $\langle 0, 1 \rangle$,

(iv) the numbers x_1, x_2, x_3 and x_4 lie each in a different quarter of the interval, etc. and, finally

(v) the numbers $x_1, x_2, \ldots, x_{10}$ lie each in a different tenth of the closed interval $\langle 0, 1 \rangle$.

7. Generalization

Is the above problem solvable if instead of 10 numbers and 10 conditions there are n numbers and n conditions (where n is an arbitrary natural number)?

8. Proportions

Consider numbers A, B, C, p, q, r whose mutual dependence is expressed as follows:

$$A:B = p, \quad B:C = q, \quad C:A = r.$$

Write the proportion

$$A:B:C = \square:\square:\square$$

is such a way that in the empty squares there will appear expressions constructed from p, q and r, these expressions being obtained from one another by a cyclic permutation of p, q and r. (By the above we mean the following: if instead of p we write q, instead of q we write r, and instead of r we write p, then the first expression $\square$ will become the second, the second expression will become the third, and the third expression will become the first.)

9. Irrationality of the root

Give an elementary proof that the positive root of the equation

$$x^5 + x = 10$$

is irrational.

10. Inequality

Prove the inequality

$$\frac{A+a+B+b}{A+a+B+b+c+r} + \frac{B+b+C+c}{B+b+C+c+a+r}$$
$$> \frac{C+c+A+a}{C+c+A+a+b+r},$$

in which all letters denote positive numbers.

11. A sequence of numbers

Find a sequence a_0, a_1, a_2, ... whose elements are positive and such that $a_0 = 1$ and $a_n - a_{n+1} = a_{n+2}$ for $n = 0, 1, 2, ...$ Show that there is only one such sequence.

CHAPTER II

PROBLEMS ON POINTS, POLYGONS, CIRCLES AND ELLIPSES

12. Points in a plane

Consider several points lying in a plane. We connect each point to the nearest point by means of a straight line. Since we assume all distances to be different, there is no doubt as to which point is the nearest one. Prove that the resulting figure does not contain any closed polygons or intersecting segments.

13. Examination of an angle

Let x_1, x_2, ..., x_n be positive numbers. We choose in a plane a ray OX, and we lay off on it a segment $OP_1 = x_1$. Then we draw a segment $P_1P_2 = x_2$ perpendicular to OP_1 and next a segment $P_2P_3 = x_3$ perpendicular to OP_2. We continue in this way up to $P_{n-1}P_n = x_n$. The right angles .are directed in such a way that their left arms pass through O. We can consider the ray OX to rotate around O from the initial position through points P_1, P_2, ... up to the final position OP_n; in doing so it sweeps out a certain angle. Prove that for given numbers x_i this angle is smallest when the numbers x_i decrease (i.e., $x_1 \geqq x_2 \geqq ... \geqq x_n$), and largest when the numbers increase (i.e., $x_1 \leqq x_2 \leqq ... \leqq x_n$).

14. Area of a triangle

Prove, without the help of trigonometry, that in a triangle with one angle $A = 60°$ the area S of the triangle is given by the formula

$$S = \frac{\sqrt{3}}{4}[a^2 - (b-c)^2];\qquad (1)$$

and if $A = 120°$, then

$$S = \frac{\sqrt{3}}{12}[a^2 - (b-c)^2].\qquad (2)$$

15. Triple halving of the perimeter of a triangle

Consider an arbitrary triangle. We can, of course, cut the triangle with a straight line to halve its perimeter. We can even impose the direction of the cutting line. If we repeat this operation twice, using two different directions, the straight lines will intersect at a certain point Q. Then, two straight lines halving the perimeter will pass through the point Q.

Does there exist a point through which three such lines pass? If so, how can we find it?

16. Division of a triangle

Divide a triangle into 19 triangles in such a way that at each vertex of the newly formed figure (and also at the vertices of the original triangle) the same number of sides meet.

In this problem the number 19 cannot be replaced by a larger number, but it can be replaced by smaller ones. Which ones?

17. Triangles

In this problem n denotes a natural number. We have $3n$ points in a plane, no three of which lie on a straight line. Can we form from these points — taking them as vertices — n triangles which do not overlap and do not embrace one another?

A similar problem for $4n$ points can be formulated for quadrangles, and in the case of $5n$ points for pentagons, etc. Are they all positively soluble?

18. Triangular network (I)

It is well known how to cover the whole plane with a network of equilateral triangles.

Is it possible to put at each of the nodes one of the signs *plus* and *minus* so that in each of the component triangles of the network the following condition will be satisfied: if two vertices of a triangle have the same sign, then the third sign is plus, and when the signs are opposite, then the third is minus?

Obviously, of course, plus signs can be given everywhere, but we exclude this trivial solution.

19. Triangular network (II)

Prove that it is impossible to cover the whole plane with a network of triangles in such a way that at every vertex five triangles would meet.

20. What is left from a rectangle?

A *golden rectangle* is a rectangle whose sides have the golden ratio, i.e. they fulfill

$$a : b = b : (a-b).$$

Let us imagine that this rectangle is cut out from a piece of paper and laid on the table with the longer side turned toward us. We cut from the left side of the rectangle the largest possible square, and what remains is again a golden rectangle. We go over to the left side of the table to have again the longer side before us, and we do the same with the new rectangle. In this way we go clockwise around the table, and we cut out the consecutive squares. Every point of the rectangle, with only one exception, will sooner or later be cut out. Find the location of this exceptional point.

21. Division of a square

Let us divide a square of area 1 km^2 into three parts A, B, C. Whatever this division may be, there always exists at least one pair of points P, Q belonging both to the same part, the distance PQ being greater than $1 \cdot 00778 \text{ km}$.

How can we prove it?

22. Square network

We can cover the whole plane with a network of congruent squares. The nodes of this network are called by mathematicians the *lattice of integral numbers*. Is it possible to mark the nodes by letters a, b, c, d so that each component square has all four letters at its vertices, and so that four letters appear in every row and column of the lattice?

23. Lattice points

For the definition of a lattice we refer the reader to problem 22. Prove that a circle with centre $(\sqrt{2}, \sqrt{3})$ may, by a suitable choice of the radius, be made to pass through any point of the lattice, but that there are no circles with this centre passing through two or more points of the lattice.

24. Lattice points inside a circle

In this problem we shall deal with lattice points inside a circle K, that is, with points enclosed by the circle K. We do not include here the lattice points on the circle itself.

Prove that there exist circles containing zero lattice points, one lattice point, two lattice points, etc. We can associate with every number n (natural or zero) a circle containing exactly n points.

25. $14 = 15$

In Wrocław in 1952, during a meeting of the participants of the Mathematical Olympiad, Dr. J. Mikusiński demonstrated a division of the whole plane into heptagons in such a way that at every vertex three heptagons meet. From this we shall deduce that $14 = 15$. Let us denote by P the angle $180°$. The sum of the angles in a heptagon is $5P$; thus the average size of an angle in a heptagon is $5P/7$. As the whole plane is covered with heptagons, it follows that the average angle in this mosaic is $5P/7$. But at each vertex three such angles meet, whence the average size of an angle at every vertex is $2P/3$. From this it follows that the average size of an angle in the mosaic is $2P/3$, because every angle belongs to a certain vertex; thus $2P/3 = 5P/7$, $2/3 = 5/7$, $14 = 15$, which was to be proved.

Find the error in the above argument.

26. Polygon

There are n points lying in a plane, no three of them lying on the same straight line. Is it always possible to find a closed polygon with n non-intersecting sides whose vertices are these n points?

27. Points and a circle

We have 4 points in a plane which do not lie on a common circle or a common straight line. Is it always possible to mark these points A, B, C, D in such a way that the circle passing through the points A, B, C contains the point D?

28. Geometrical problem

Given an ellipse with major axis of length $2a$, and minor axis of length $2b$, draw a closed curve of the same length as the length of the ellipse such that the closed curve encloses an area greater than the area of the ellipse by $(a-b)^2$.

CHAPTER III

PROBLEMS ON SPACE, POLYHEDRA AND SPHERES

29. Division of space

Through a fixed point in space we pass planes to divide the space into the greatest possible number of parts. One plane will divide the space into two parts, two intersecting planes into four, and three planes intersecting at the fixed point and having no other common points divide the space into eight parts. What is the number of parts obtained with four planes? What number do we get in the case of n planes?

30. Two projections

Let us imagine a plane Π_1, tangent to the globe at the north pole N, and a plane Π_2 tangent to the globe at the south pole S. We can draw a map by projecting each point on the surface of the globe from N to Π_2, and another map by projecting each point on the surface of the globe from S to Π_1; these are the so-called *stereographic projections*. Now, we can superimpose the two planes on each other so that the meridians agree. Each town on one map appears also on the second one. Thus we have defined a certain transformation of the plane into itself. How can we define this transformation directly?

31. Cube

Holding in the hand a model of a cube so that it can rotate about the longest axis (i.e. around the line joining two opposite vertices), we can wind black yarn closely around the cube. This yarn will cover only half of this cube (why?). The same can be done with other axes; there are four of them, and each time we take a yarn of diffcrent colour (black, red, blue and yellow). The whole cube will be covered with overlapping colours, producing mixed shades (the model of the cube is white, which we do not consider as a colour). How many shades will result and which ones?

32. Geodesics

This problem does not require much mathematics. Let us place on a smooth rigid cube a rubber band, such as are used in chemists' shops to fasten packages, in such a way that it stays on the cube but does not cross itself. We call the line defined by the rubber band a *geodesic* and we draw all such geodesics on our model.

1. How many times will the geodesics cover the surface of the cube (that is, how many geodesics pass through each point of the surface of the cube)?

2. How many different families of geodesics have the property of covering the surface of the cube?

33. Motion of a particle

Inside a cubic box there is moving a material particle free from the action of external forces; it is reflected from the walls of the box according to the classical law (the angle of reflection is equal to the angle of incidence, and the perpendicular to the wall at the reflection point is the symmetry line of the angle made by the path of the particle hitting the wall). Is it possible that such a particle runs incessantly around a closed hexagon, hitting on every trip all the walls of the box in turn? Define the points of reflection, and examine whether or not this hexagon is knotted.

34. Diagrams of the cube

Models of polyhedra are made from flat diagrams drawn on pasteboard. On the diagram the faces are adjacent, and one makes a model of it by folding the pasteboard along the lines of the diagram. A regular tetrahedron has two different diagrams.

How many has the cube?

35. Cubes

As we know, the whole of space can be filled with cubes adjoining one another. At every vertex eight cubes meet. It is possible, therefore, by cutting off the corners of those cubes in a suitable way and glueing together the eight adjacent fragments thus obtained (which gives us a new polyhedron and a regular octahedron) to fill the space with regular octahedra, and with the solids that remain of the original cubes. What kinds of solids will these be? If we enlarge the octahedra as much as possible, what part of the space will they occupy? What will be the size of the remaining solids? And, how many solids will meet at every vertex?

36. Hexahedron

Does there exist a hexahedron which is limited by congruent rhombi but is not a cube in the usual sense, i.e. not a die?

37. Tetrahedra

We have six sticks of different lengths, and they are such that in every permutation they can be considered to be edges of a tetrahedron. How many tetrahedra can be constructed from these sticks?

38. Tetrahedron with congruent faces

Is it possible to construct a tetrahedron whose faces are mutually congruent triangles with sides of arbitrary lengths a, b and c?

If so, compute the volume of the tetrahedron.

39. Octahedron

Is it possible to construct an octahedron whose faces are congruent quadrangles? Is it possible to construct a decahedron, and in general, to construct a $2n$-hedron $(n > 3)$ with the same property?

40. Distance on a surface

On a closed and convex surface we can associate with every pair of points an arc which joins them and is the shortest of all such arcs. This does not exclude the possibility of there being another arc which also has the least length; for example, on the sphere any pair of antipodes admits of infinitely many shortest connections. If, for any pair of points A, B on the surface, by "the

distance *AB"* we mean the length of the shortest arc *AB*, then
we shall be able to speak about the distance *PX* between *P* and a
point *X* arbitrarily given on the surface. Because of this we shall
be able to associate with the point *P* a point *furthest* from it; let
us call it *Q*. (There can be more than one such furthest point.)
One could suppose that such a pair *P*, *Q* always has at least two
shortest connecting arcs. Show that this supposition is false for
certain tetrahedra.

41. The wandering of a fly

A fly sat down at a vertex of a regular dodecahedron, and decided
to visit vertices walking along its edges in such a way as not to

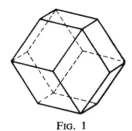

Fig. 1

pass by any vertex twice, and to return to the starting point. Having done so it tried again on a rhombic dodecahedron (Fig. 1).
Was the trial succesful?

42. Regular dodecahedron

A regular dodecahedron can be painted with four colours in
such a way that the adjacent faces are all differently coloured.
Prove that there are only four ways of solving this problem, if
we consider as the same solution two models painted in such a way
that by rotation we can reach the same distribution of colours
on both models.

43. Polyhedron

A convex polyhedron must have convex faces. Conversely,
must a polyhedron with convex faces be convex itself? In particular,
do there exist two polyhedra (for instance, two 30-hedra) bounded

by the same number of faces coinciding pair-wise (faces of the same figure need not be mutually congruent), of which one is convex and the second is not?

44. Non-convex polyhedron

Can a non-convex polyhedron be bounded by congruent quadrangles?

45. Problem from Wonderland

Lewis Carroll, a mathematician and writer of children's stories, was a creator of wonderful absurdities. He advised, for instance, the use of a map in the ratio 1:1, because it suffices to unfold it on the earth to know at any instant where one is located; quite simply: you read the inscription on which you are standing.

Let us imagine that, following this advice, we paint the meridians and parallels on the globe with a lasting colour, and that we put on it calligraphic names of towns, ports, and countries. A compass will not be necessary, but one difficulty remains: how to find the shortest way to the goal? We know that *orthodromes* (which mean the shortest ways) are not *loxodromes* on this map of Alice-in-Wonderland, i.e. lines crossing the meridians and parallels at a constant angle. What is worse, no repainting will help: all systems of orienting lines will have this shortcoming. The fault, of course, belongs to our globe, which is inconveniently constructed.

The best way to reform the globe is to begin with the map. It is possible, for instance, to draw a rectangular diagram of meridians and parallels and roll this map into a cylinder, so that parallels become circles. On such a cylindrical planet the shortest route from point to point always crosses a meridian at a constant angle. It is possible also to cut the map along a parallel, mark a point N on it, and roll it into a cone with apex N. This conical planet will have the point N as the north pole, the parallels will not cross one another, and neither will the meridians; but each parallel will intersect each meridian at two points — just as on the globe. As before, the shortest distances will again have a constant course.

But it is possible to find a still more interesting model. On the map there will be a rectangular network of orientating lines, but there will appear on the planet only one family of lines: every line will intersect every other line at two points, and also itself at one point. The principle of the constant course will remain.

What is this model?

Mr. R. Nowakowski, when I related both models to him, immediately found a third one: a rectangular network consisting of meridians, parallels, and "merillels".

46. Three spheres and a line

Three spheres have a common point P, but no line through P is tangent to all three of them. Show that these three spheres have an additional common point.

47. A property of the sphere

Let all plane sections of a certain surface be circles (a single point is taken to be a circle of radius 0). Show that this surface is a sphere.

CHAPTER IV

PRACTICAL AND NON-PRACTICAL PROBLEMS

48. Puzzle

On a cardboard disk a smaller disk is marked by a circle concentric with the boundary of the cardboard. The smaller disk is divided into 8 sections of the same size: 4 white and 4 black. The remaining ring is divided into 10 sections, white and black in turn, 5 of each colour. The toy is stuck on a nail, and rotated very rapidly. At first the colours mix into a uniform greyness. After a moment the ring appears to be rotating in one direction, and the smaller disk rotating in the opposite direction, although the whole toy consists of one piece of cardboard.

This toy functions under electric light only, and not in every town. Why?

49. Picnic ham

Three neighbours gave $ 4·00 each and bought a ham (without skin, fat, and bones). One of them divided it into three parts asserting that the weights were equal. The second neighbour declared that she trusted only the balance of the shop at the corner. There, it appeared that the parts, supposed to be equal, corresponded to the monetary values of $ 3·00, $ 4·00 and $ 5·00, respectively. The third partner decided to weigh the ham on her home balance, which gave a still different result. This led to a quarrel, because the first woman kept insisting on the equality of her division, the second one recognized only the balance of the shop, and the third only her own balance. In what way is it possible to settle this dispute and to divide these pieces (without cutting them anew) in such a way that each woman would have to admit that she had got at least $ 4·00 worth of ham if computed according to the balance which she trusted?

50. Quartering of a pie

Every pie, regardless of its shape, can be quartered evenly by two mutually perpendicular cuts. In other words, for every plane domain of area P it is possible to find two perpendicular lines such that in each of the 4 quarters made by them there lies a part of the domain, of area $P/4$. Prove this assertion.

(The proof of the general theorem is easier than the effective quartering of a triangular pie with sides 3, 4 and 5.)

51. Another pie

Ted and Ned are to share a triangular pie. Ned imposes the condition that he will cut his part with one straight cut, and Ted agrees; however, Ted requests the privilege of determining the point P through which Ned's line will pass. As the pie has equal thickness everywhere, and is also uniform as to taste, then the difficulty reduces to a question of plane geometry. This question is: In what way can Ted mark the point P to best protect himself against the greediness of Ned? The second question: How large a surplus will Ned obtain if Ted solves the first problem correctly and Ned afterwards takes for himself the largest possible part of the pie?

If the shape of the pie depended on Ted, then he would be able to choose a figure having a centre (circle, square, ellipse etc.), and put P there. In this case Ned's privilege would become meaningless.

But there is a more interesting question — what shape of the pie (if we retain the conditions of division stated above) will be the best for Ned, and what is the largest surplus Ned can ensure for himself by the proper choice of the shape?

52. Weighings

We have 5 objects which differ in weight, and we wish to arrange them in a sequence of decreasing weights. We possess primitive scales, without a set of weights, on which we can compare the objects pairwise. How must we proceed in order to arrange the objects in the fastest possible way? (That is, to reduce the number of comparisons to a minimum.) How many comparisons will there be?

53. How old is Mrs. Z?

Mrs. Z is not very old, as she was born after World War I, but she does not like to answer directly questions concerning her age.

Asked such a question on 27 July 1950, she answered: I am only one year old, as I celebrate my birthdays only when they fall on the right day of the week, and I have had only one such case in my life.

54. How many fish are there in the pond?

An ichthyologist wanted to estimate the number of fish in a pond which are suitable to be caught. He threw into the pond a net with regulation size mesh, and after having taken the net out he found 30 fish in the net; he marked each of them with a suitable colour, and threw them all back into the pond. The next day he threw the same net and captured 40 fish, of which 2 had been marked. In what way did he compute (approximately) the number of fish in the pond?

55. Calibration of rollers

One part of a petrol engine has the shape of a roller. To measure its diameter a steel plate is used in which there is a row of 15 holes, of sizes determined in a precise way. The first hole has a diameter of 10 mm, and each succeeding one is 0·04 mm larger in diameter than the preceding one. The calibration of the roller consists in fitting it into the holes; a hole is selected (arbitrarily), and the size of the roller is tested. If the roller cannot enter the hole we consider its diameter to exceed the diameter of the hole. If it does go in we consider its diameter to be smaller. In this way we eventually determine the diameter of the roller with error less than 0·04 mm; rollers with diameters less than 10 mm, or greater than 10·56 mm are discarded, and the remaining ones are sent to the next stage of treatment.

The worker entrusted with the calibration tries each roller in the same number of holes, though he does not use the same holes. How many tests does each roller require? What is to be the order of testing?

56. 120 ball-bearings

A precision workshop ordered 120 ball-bearings of diameter 6·1 mm. 120 ball-bearings were supplied, but accurate measurements showed that the diameter did not meet the required specification. Namely, there were

10 ball-bearings with diameter 6·01 mm
6 „ „ „ „ 6·02 „
4 „ „ „ „ 6·03 „
10 „ „ „ „ 6·05 „
19 „ „ „ „ 6·07 „
11 „ „ „ „ 6·08 „
6 „ „ „ „ 6·10 „
6 „ „ „ „ 6·11 „
8 „ „ „ „ 6·12 „
10 „ „ „ „ 6·14 „
17 „ „ „ „ 6·16 „
6 „ „ „ „ 6·17 „
7 „ „ „ „ 6·18 „

Fortunately, another workshop agreed to accept the ball-bearings under the condition that they be supplied in two boxes — one containing the larger ones, and one containing the smaller ones, and that on each box the average diameter be given. The problem is to give the critical diameter d below which the ball-bearings will be put in box A (thus box B will contain all ball-bearings with diameter larger than the critical diameter), and to find numbers a and b which ought to be written as signs on the boxes. These three numbers a, b, d are to be such that the sum of the absolute errors is minimized. Here, the absolute error means the absolute difference between the diameter of the ball and the number on the box into which it was put.

57. Ribbon on the roll

25 metres of ribbon, 0·1 mm thick, was closely wound on a roll made from pasteboard, giving a new roll of diameter 1 decimeter. What is the diameter of the pasteboard roll?

58. Watch with both hands identical

We know that in defining the time without a watch nobody makes a mistake greater than six hours.

A watchmaker put on a watch two hands of equal length, so that it was impossible to distinguish the minute hand from the hour hand. What is the maximum error that the owner of the watch can make?

59. Problems of giants and midgets

During a gymnastics class, in which all the pupils were of different heights, the teacher arranged the pupils in a rectangular order, and said: "Now we shall see who is the tallest midget". He found the shortest pupil in every column, and when the midgets had stepped forward and formed a row, he chose the tallest of this group and said: "Here is the tallest midget".

After the boys then returned to their former places the teacher said: "Now let me see who is the shortest giant". He pointed to the tallest boy in each row, and when these giants formed a flank, he found the shortest one and said: "Here is the shortest giant".

Is it possible that the same boy was the tallest midget and the shortest giant? Do there exist classes in which the shortest giant is smaller than the tallest midget? And what would be the result if the teacher looked for the giants in columns and not in rows, that is, in the same way as he looked for midgets?

60. Acks and backs

A large class of pupils is arbitrarily divided into two sections. We call the pupils from section *A* "acks" and those from section *B* "backs". The acks boast of being taller than the backs; and the backs enjoy the opinion of being better mathematicians. Once when one of the acks looked down upon one of the backs, the back asked: "What does it mean that you are taller than we are? Does it mean that:

1. Each ack is taller than each back?
2. The tallest ack is taller than the tallest back?
3. Each ack is taller than some back?
4. Each back is smaller than some ack?

5. Each ack has a corresponding back (and each of them a different one) whom he surpasses in height?

6. Each back has a corresponding ack (and each of them a different one) by whom he is surpassed?

7. The shortest back is shorter than the shortest ack?

8. The shortest ack exceeds more backs than the tallest back exceeds acks?

9. The sum of the heights of the acks is greater than the sum of the heights of the backs?

10. The average height of the acks is greater than the average height of the backs?

11. There are more acks who exceed some back than there are backs who exceed some ack?

12. There are more acks with height greater than the average height of the backs than there are backs with height greater than the average height of the acks?

13. The central height of the acks is greater than that of the backs (in the case where the number of pupils in the class is even, we take as the central height the arithmetic mean of the heights of the central pair of pupils)?"

Flooded with this torrent of questions the ack dwindled in size! We ask the reader: Which of the 13 questions are independent of each other, and which of them are mutually dependent? In other words, find all pairs of questions such that the answer "yes" to the first one implies the answer "yes" to the second one. Are there any questions that are equivalent; that is, are there any pairs such that the answer to both questions has to be the same? Are there any pairs which are dependent but not equivalent?

61. Statistics

A statistician decided to examine the utilization of compartments for smokers and non-smokers in trains in different countries. He distinguished the following possibilities:

(*a*) smokers mostly go into compartments for smokers,

(*a'*) non-*a* ("non-*a*" denotes the opposite of *a*),

(*b*) non-smokers mostly go into compartments for smokers,

(*b'*) non-*b*,

(c) compartments for smokers are mostly used by smokers,

(c') non-c,

(d) compartments for non-smokers are mostly used by smokers,

(d') non-d.

Each country can be characterized by four letters a, b, c, d, with or without a prime; of course no letter can appear both with and without its prime, as each primed proposition is the negation of the unprimed one. Thus there are sixteen symbols.

Is it possible to arrange people in sixteen trains so that each train has a different symbol?

62. Blood groups

It is well known (Landsteiner, Jański, Moss, and others) that all people can be divided into four blood groups: O, A, B, AB (terminology of Dungern and Hirszfeld; this classification enables us to decide whether an individual can give his blood by transfusion without danger to the receiver). Let us denote (symbolically) by $X \to Y$ the statement: "an individual belonging to group X can always give his blood to an individual belonging to group Y without danger to the latter". Then the laws discovered by the scientists mentioned above can be formulated as follows:

I. $X \to X$ for each X.

II. $O \to X$ for each X.

III. $X \to AB$ for each X.

IV. Any relation $X \to Y$ which does not result from I, II, III by substitution of the symbols O, A, B, AB for X is false.

Prove that

(1) The system of laws I–IV is not contradictory;

(2) Assuming the validity of laws I–IV, it follows from $X \to Y$ and $Y \to Z$ that $X \to Z$ for all X, Y, Z;

(3) I–IV imply non-$(A \to B)$.

Explanation. The expression "for each X" in I, II, III means that the implications $\to$ are valid for X equal to O, A, B, AB. A similar remark applies to (2).

63. Blood groups again

Felix Bernstein (whose name is associated with the theory of sets) was the first to formulate the laws of inheritance for blood groups O, A, B, AB. Suppose, for instance, that the father belongs to group A and the mother to group AB. Let us ascribe to the one-letter symbol A the letter O; i.e. we call the group to which the father belongs AO. The groups of both parents will thus be AO, AB. To formulate the symbol for the child we must take one letter from the mother and one from the father. We thus obtain AA, AB, OA, OB as possible symbols for the children.

These symbols are then simplified by writing instead of the double letter AA the single letter A, and omitting one O whenever it occurs in a two-letter symbol. Thus we obtain A, AB, A, B. Consequently, in our example, the child can belong to any of the three groups A, B, AB, and cannot belong to O.

Thus the rules of ascribing O, taking one letter from either parent, the reduction of double-letter symbols, and omitting O, define entirely (and not only in the example cited above) the so-called phenotypic theory of inheritance of blood groups. We gave the laws of transfusion in problem 62.

Two brothers knew the laws of transfusion and knew that neither of them would be able to give his blood to the other one, but each of them could be given the blood of their mother. Would their sister be able to replace the mother?

64. Excess of labour

When someone wants to hammer a nail in each of several posts, put at equal distances along a road, the best way is to begin with the first post and finish with the last one. But how can we accomplish this task in the worst way, that is such that the route be the longest?

65. Diagonal of a wooden block

Having a ruler at our disposal we can measure the diagonal of a wooden block, that is, the distance between the most distant corners.

Give a practical way to measure the diagonal, such as might be used in a workshop (and not a scholarly application of the theorem of Pythagoras).

66. The tying of boxes

In sweet shops, assistants tie boxes of sweets as follows: The ribbon runs diagonally around the box, and forms one closed skew eight-sided figure. Both on the lid and on the bottom two parallel segments of the ribbon can be seen. Knowing the dimensions of the box, show that it is possible to compute the length of the ribbon, as well as the angles at which the ribbon cuts the sides. Finally, prove that this ribbon can not only be pushed along itself, but also displaced on the box.

67. A primitive device

A very simple balance can be made as follows: It consists of a wooden stick of constant thickness, made of uniform material, which passes at one end into a heavy block; at the other it has a hook upon which objects are hung in order to be weighed. There are notches on the stick forming a scale on which we can read the weight (in pounds, say) of any object which hangs on the hook. To do this, it is necessary to find the place at which the stick, supported by the finger (or blade of a knife), is balanced. The number read on the scale gives the weight sought. It is easy to make such a scale experimentally when one possesses a set of standard weights. The more weights available, the greater the precision of the instrument.

How can we construct such a scale geometrically if there is only one standard weight (say of one pound) available?

68. The minimal length

A ruler L is fixed to a table, and another ruler R, resting constantly with its edge against a nail O hammered into the table, glides with its corner B along the edge of ruler L (Fig. 86). The edge which rests against O ends at the corner A. By such a motion the distance AO has its minimum in a certain position of the ruler R.

Define this position, and compute the minimum AO, knowing the distance of the nail from the fixed ruler and the width of the mobile ruler.

69. Division of plots

As is well known, plots of land usually have rectangular shapes. Let us imagine that all plots are rectangular. We also know that successive partitions of plots result from the division among the heirs of the deceased owner. When we see a rectangular plot and also see a boundary line which again divides the plot into two rectangular lots, then it is evident that this configuration is the result of one division, and it could not have resulted otherwise. But, if we divide the original plot into three rectangular plots, then someone who does not know the history of this land (directly or from the mortgage records) would not be able to decide if this configuration resulted by an original division among three heirs, or by a division into two plots, and then by a later division, in the next generation, of one of these plots into two plots. We say that the division into two parts is *primary*, and the division into three parts is not primary. More clearly, by a primary division we mean a division that could not have arisen from successive divisions (it is quite immaterial how it has arisen). This definition has been given by Dr. J. Łoś, who has observed that there exist primary divisions into 2, 5, 7, 8, ..., etc. parts, but there are no primary divisions into 3, 4 and 6 parts. (The reader will prove for himself that there are no primary divisions into 3 and 4 parts, and he will find primary divisions into 5 and 7 parts. Dr. Cz. Ryll-Nardzewski has proved that there is no primary division into 6 parts.)

(1) Give a primary division of a square plot into 5 equal parts.
(2) Give a primary division of a square plot into 7 equal parts.
(3) Give a primary division of a square plot into 8 equal parts.

70. A practical problem

The land under a factory is flat, but inclined. We have a device to determine the level of the land; it consists of a horizontal telescope which rotates about its vertical axis (the angles of rotation being read on the scale on a horizontal circle) and a stick at which we aim our telescope in order to read the difference of the levels, and at the same time the distance to the stick. What is the simplest possible way to find the inclination of the land and the direction of the inclination?

71. Neighbouring towns

On the map of Europe we connect every town with the nearest one assuming that the distances between them are never equal. Prove that no town will be connected with more than five neighbouring towns.

72. Railway lines (I)

There are five towns no three of which lie on a straight line. They must be connected by a railway line composed of four straight tracks. Bridge crossings are permitted; the lines are supposed to have different levels at the crossing points.

How many different railway lines can be constructed?

73. Railway lines (II)

Cities A, B, C, D lie at the corners of a square with sides of 100 km. We must plan the railway lines so that each city is joined with each of the remaining three cities (junction stations other than cities A, B, C, D are permitted), and the total length of the lines has to be minimized. What is the system of railway lines we seek, and what is the total length of the lines?

74. Test flight

A jet plane of a new type started in Oslo and took the shortest route to an airport X in South America lying right on the Equator. The witnesses of the departure in Oslo saw the plane disappear on the horizon at a point lying almost due West.

How long is the route of flight? At which point on the horizon should the spectators waiting at airport X expect the plane?

75. Sun and Moon

The distance of the Sun from the Earth is 387 times the distance of the Moon from the Earth. How many times does the volume of the Sun exceed that of the Moon?

76. Cosmography

Compute the length of the shortest day in Wrocław, Poland. The solution requires the knowledge of two angles. Which ones?

CHAPTER V

PROBLEMS ON CHESS, VOLLEYBALL AND PURSUIT

77. Chessboard

Consider a square or rectangular chessboard with an odd number of squares (e.g. 49 or 63). Squares with common sides are called *adjacent*.

We place a pawn on each square, then we pick up the pawns and with them cover again the whole chessboard at random.

Is it possible for every pawn to find itself on a square adjacent to its original position?

78. Chessboard revisited

We place one pawn on each square of the chessboard. We pick up these pawns and place them again, but in such a way that the pawns that stood in the left-hand corners take their original positions, and pawns that were next to one another (i.e. that occupied adjacent squares) are again neighbours.

Is it possible for any pawn to occupy now a different position from its old one?

79. Rooks on the chessboard

The chessboard we shall consider has as many rows as columns, but differs from the usual one in the distribution of the white and black squares, which is arbitrary with the following restrictions: every column has at least one white square, and at least one column is entirely white. We shall say that we have succeeded in placing the rooks on the chessboard (we have a sufficient number of pieces not to run short of them), if we satisfy the following conditions: (1) the rooks are placed only on the white squares, (2) at least one rook is placed on the chessboard, (3) the rooks do not attack each other (i.e. they are not standing in such a way as to be able to capture each other), (4) every white square not occupied by a rook, but attacked horizontally by a rook, is also attacked vertically by some rook. Prove that it is always possible to place the rooks according to conditions (1), (2), (3) and (4).

80. Elliptical billiards

On an elliptical billiard table ball A is standing by the cushion, and ball B on the segment S connecting the foci of the ellipse. We are to strike A in such a way that, rebounding from the cushion, it hits B. However, it is forbidd in for A to cross the segment S before touching the cushion.

Show that this problem cannot be solved.

81. A sports problem (I)

There are 25 pupils in a class. Among them 17 pupils can ride a bicycle, 13 can swim and 8 can ski. No pupil can perform all three sports, but the cyclists, swimmers, and skiers have all received grades B or C in mathematics. This is interesting since 6 pupils in the class have received grades of D, E, or F (no credit) in mathematics. How many pupils have received grade A? How many swimmers can ski?

82. A sports problem (II)

Three runners, A, B and C, systematically trained together for the 220 yards race, noting after every practice run the order in which they reached the finish. When the track season was over they found that in the majority of informal (or practice) competitions A surpassed B, and that in most runs B appeared to be faster than C, and also in the majority of trials C performed better than A. How could this happen?

83. Theory of sport eliminations

The Dr. Sylvester Abracadabrus Chess Club has 10 members. Every year a competition takes place in order to divide the players into classes. Each member plays with every other member, and plays until check mate results.

We shall say that "A beats B" if A has beaten B in this year's elimination competition. When the competition is finished there will be 45 such results, and the players will be divided into classes, for instance of those members who have beaten 8 others, of those who have beaten 7 others, etc. (Let us remark that this system of 45 competitions leads to the possibility of A beating B, B beating C, and C beating A.)

The question concerns the possible classifying results. In particular, is it possible that the club should split into 3 classes?

84. Volleyball league

The best volleyball teams form a league which arranges meetings during the season; each team plays against every other team once. It can happen that one of the teams will beat the others, but this need not always be the case. Then let us agree to call "leader" the team which beats every other team either in the usual way or indirectly through a third team. In other words, we consider that A has defeated B if there is a team C such that the match between C and A was victorious for A and the match CB victorious for C. However we shall *not* consider as a victory of A over B the case where A beats C, C beats D, and D beats B.

Prove that (1) league competitions always lead to recognition of one or more leaders, (2) the team which won directly the greatest number of competitions is always a leader.

85. Tournaments

The so-called cup system of determining the master (for instance, in lawn-tennis tournaments) from among eight players consists of arranging the players in pairs by drawing lots. Four matches determine four winners, who play a second round; the third round is the final match. The winner of the final match receives the first prize, and his opponent receives the second prize. Let us imagine that each of the players has a defined strength, as every material object has its weight, and that the stronger player always beats the weaker one (just as it is always the heavier object which overbalances the lighter one if they are placed on opposite scales). If so, then the procedure described above designates the champion in a just way, because it shows that he is the strongest competitor. But, it does not always designate the second place in a just way.

What is the probability that the finalist really deserves the second prize?

86. Bicyclist and walkers

The manager of an estate sent two walking messengers: one with a letter to the post office, and the second one, a quarter of hour later, in the opposite direction to a shop. He soon realized that he had made a mistake in distributing the letters to the messengers; so he sent a bicyclist to overtake both messengers, rectify his mistake and return. The bicyclist assumes that both messengers walk at the same speed, and wonders which messenger he should pursue first. As he can ride fast, he will carry out the order of the manager in either case.

The reader who solves this problem ought to say what the solution would be if the manager did not change the letters, but only forgot to give money to the messengers, and wanted to correct this mistake.

87. Four dogs

Four dogs A, B, C and D are standing in the corners of a square meadow and suddenly begin to pursue one another as shown by the arrows below. Each dog is running directly after his neighbour — A after B, B after C, C after D, and D after A. The side

of the meadow is 100 metres, and the speed of each dog is 10 metres/sec. After what period of time will the dogs meet one another? Will their tracks ever cross, and where? How long will their tracks be?

88. Chase (I)

The ship P sights the ship Q which sails in a direction perpendicular to PQ and keeps her absolute course. P pursues Q aiming constantly at Q. The speed of both ships is the same at every moment (but it can vary in the course of time). It is evident without computation that P is sailing along a curved line. If the pursuit lasts for a long time the track of the pursuing ship becomes almost identical with the track of the escaping ship. What is the distance PQ if at the beginning it was equal to 10 nautical miles?

89. Chase (II)

The ship O_1 sights another ship O_2 sailing at right angles to the line of sight O_1O_2 at the moment of observation. O_2 does not recognize the signals of O_1, and keeps her initial course and initial speed v_2. O_1 wants to call attention to herself as she requires rescue, and she sails at the maximum speed v_1 of which she is capable, in such a direction as to approach O_2 as closely as possible. Which course must O_1 choose? What will be the final distance if the initial distance between O_1 and O_2 was d and the ratio of speeds v_1/v_2 is equal to $k < 1$?

90. Incomplete data

Someone who had read the last problem not very carefully described it to Dr. Abracadabrus, and asked him how to find the course from the data given in the problem. Unfortunately he forgot which ship had the greater speed. He did remember that the ratio k of the speeds was less than 1 and known, but he did not know whether k denotes the ratio v_1/v_2 or v_2/v_1. He was very astonished when the Master immediately found the course on the basis of incomplete data. How did Dr. Abracadabrus do it?

91. Motorboat (I)

A smugglers' motorboat has a speed three times greater than the boat of customs officials which is half-way between the smugglers' boat and the point on the coast which it wishes to reach.

The captain of the smugglers' boat decides to reach the coast by sailing along two sides of a square. Which part of this course will be dangerous?

92. Motorboat (II)

Suppose that in the last problem the captain of the smugglers' boat decides to sail in such a way as to change his course by 90° only once. Which course should he choose so as to be sure of evading the pursuing ship and reaching the coast in the shortest possible time?

MATHEMATICAL ADVENTURES OF DR. ABRACADABRUS

93. The strange number

Dr. Abracadabrus decided to reform mathematical notation entirely. He considers it to be a great scandal that there is a number known to children just beginning their education, which has besides its common symbol a second one, about which these children learn several years later, and a third one, very complicated, about which (alas!) they learn as adults; and at the same time nobody tells them that this number is the same. Only Dr. Abracadabrus and a group of his friends know this secret. What is this number?

94. The tailor's tape

Dr. Abracadabrus has a tailor's tape different (of course!) from that of ordinary mortals. It looks like the one shown in Fig. 2.

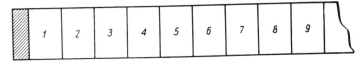

FIG. 2

At the beginning of the tape there is a border made of a 1/2 cm piece of metal. Dr. Abracadabrus states that his tape is better than the one commonly used. Why?

95. Word quiz

Dr. Sylvester Abracadabrus has stated publicly that he always wins a word quiz if he is allowed to put 20 questions, the answers to which must be "yes" or "no" (according to the rules of the game), and if the word to be guessed belongs to the dictionary. Can he substantiate his claim?

96. Student debts

Seven students are living together in one apartment. During the whole year they lend each other small amounts of money. Dr. S. Abracadabrus advised that each of them should keep a record of the amounts he lends to the others, and the amount he borrows, but not to note whom he lends to or borrows from. Before their departure for the vacation the students decided to settle their accounts, but they did not know how to do it.

Does the accounting system of Dr. Abracadabrus suffice to settle the accounts? How many payments are necessary in the worst case (we call the handing of a certain sum of money from one person to another a payment).

97. A strange social set

Someone related that he had once found himself in a social set, consisting (together with himself) of 12 persons, which was constituted in such a way that

(a) Everybody in the group was unacquainted with 6 of the other persons and knew the rest.

(b) Everybody belonged to some mutually acquainted triplet.

(c) There was no group of 4 persons in which all members knew one another.

(d) There was no group of 4 persons in which nobody knew anybody.

(e) Everybody belonged to a triplet not knowing one another.

(f) Everybody could find among the persons with whom he was unacquainted a person with whom he had no mutual acquaintance within the group.

Having heard this, Dr. Abracadabrus stated that he had been in a social set in which conditions (b), (c), (d), and (e) were satisfied, but everybody knew 6 and only 6 persons and had an acquaintance who could introduce him to the rest of the set.

Explain these facts.

98. Abacus

Let us imagine an abacus consisting of ten horizontal wires, with one bead on each wire. Let these beads move with a constant speed (equal for all beads), moving to and fro along the wire and turning when they hit the frame of the abacus. The initial positions of the beads are unknown. The perpendicular axis of symmetry divides the abacus into left and right halves. Let us assume that the motion takes place in such a way that there will never be more than seven beads in the right-hand half of the abacus. Dr. Abracadabrus states that there will never be less than three beads. Is he right?

99. Washing the streets

Although Dr. Abracadabrus's hometown has tank cars to wash the streets, it has no suitable garage for them. The city fathers handed Dr. Abracadabrus the map of the city (which is reproduced in Fig. 3) asking him to point out the best place for a garage,

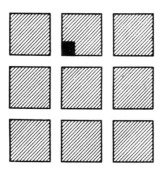

FIG. 3

that is, a place that would permit the cars to cover all streets in the shortest way and return to the garage. Dr. Abracadabrus chose his own home, which is indicated on the map by a black square.

Was he right?

100. French cities

Dr. Abracadabrus, an expert on strategy, was interested in World War II, and in 1940 he learnt the geography of the French theatre of war, Hence the following problem arose: The distance (in the sense of straight-line distance) from Châlons to Vitry is 30 km, from Vitry to Chaumont 80 km, from Chaumont to St. Quentin 236 km, from St. Quentin to Reims 86 km, from Reims to Châlons 40 km. The problem is to compute in this closed polygon the distance from Reims to Chaumont. Only Dr. Abracadabrus can do it without a map!

CHAPTER VII

PROBLEMS WITHOUT SOLUTION

The solutions of some of the problems here are known, but the majority of the problems in this chapter have not as yet been solved. In this chapter there are some easy and some difficult problems, but the solution of any of them can testify to the ability of independent thinking.

Plus aud minus sings

The figure given below consists of 14 plus signs and 14 minus signs. They are arranged in such a way that under each pair of equal signs there appears a positive sign and under opposite signs a minus sign.

```
+ + − + − + +
 + − − − − +
  − + + + −
   − + + −
    − + −
     − −
      +
```

If the first row had n signs, then in an analogous figure there would be $\frac{1}{2}n(n+1)$ signs; our example corresponds to the case $n = 7$. As $\frac{1}{2}n(n+1)$ is an even number for $n = 3, 4, 7, 8, 11, 12, \ldots$, etc., we can ask whether it is possible to construct a figure analogous to the above one and beginning with n signs in the highest row. In particular we ask about the case $n = 12$.

The general solution is not known. Here is a solution for $n = 12$ and $n = 20$, from which we obtain the solution for $n = 11$ and $n = 19$ by rejecting the first row.

```
          − + + − − + − − + + + −
           − + − + − − + − + + −
            − − − − + − − − + −
             + + + − − + + − − −
              + + − + − + − +
               + − − − − − −
                − + + + + +
                 − + + + +
                  − + + +
                   − + +
                    − +
                     −
```

```
 + + + + − − + − − − + + + + − − − − − +
  + + + − + − − + + − + + + − + + + + −
   + + − − − + − + − − + + − − + + + −
    + − + + − − − − + − + − + − + + −
     − − + − + + + − − − − − − − + −
      + − − − + + − + + + + + + − −
       − + + − + − − + + + + + + − +
        − + − − − + − + + + + + − −
         − − + + − − − + + + − +
          + − + − + + − + + − −
           − − − − + − − + − +
            + + + − − + − − − −
             + + − + − − − + +
              + − − − + − +
               − + + − − −
                − + − + +
                 − − − +
                  + + −
                   + −
                    −
```

Triangle in a triangle

A triangle has sides a, b, c, another triangle has sides a', b', c'. What conditions connecting numbers a, b, c and a', b', c' are necessary and sufficient to put the first triangle inside the second one?

Parts of a square

Let the unit square be divided into three arbitrary parts. Let r be any number less than $\sqrt{65/64}$. Is it always true that there is one part containing at least two points P, Q with $PQ = r$?

Division of a circle

We mark a point P on a circle of perimeter 1, and we cut off from P consecutive arcs of length v, v being an irrational number. In this way we get the points P_1, P_2 etc. and the length of the arc $P_k P_{k+1}$ is always v. If we stop at the point P_{n-1} we see the circle divided into n parts. Show that the point P_n will fall in the longest of these parts.

Radii in space

Three radii in space originate from a common point and form plane angles a, b, c. Which of the inequalities (1)–(5) given below are generally true and which are not?

(1) $a+b.> c$,

(2) $\sin a + \sin b > \sin c$,

(3) $\sin\frac{1}{2}a + \sin\frac{1}{2}b > \sin\frac{1}{2}c$,

(4) $\sin^2 a + \sin^2 b > \sin^2 c$,

(5) $\sin^2\frac{1}{2}a + \sin^2\frac{1}{2}b > \sin^2\frac{1}{2}c$.

Unlimited chessboard

On an unlimited chessboard we have to mark off a figure consisting of 100 unit squares, i.e. 100 squares of the chessboard, in such a way that the diameter of the resulting figure should be as small as possible. We call the greatest of all distances between points of a figure its *diameter*.

Give this diameter. Give the radius of the smallest circle enclosing 100 squares of the chessboard.

The abacus again

Let the beads spoken of in 98 carry letters a, b, c, d, e, f, g, h, i, j. In every position the projections of these beads on the base of the abacus give one of the possible permutations of the letters a, b, ..., j. Let these beads pass along the wires as in problem 98, each of them with a constant speed, the speeds being all different. Every speed is expressed by an integral number of cm/sec. Is it possible to choose the speeds in such a way that, whatever the initial position of the beads, they exhibit during their movement all the possible permutations of the letters $a, b, ..., j$? Dr. S. Abracadabrus asserts that he knows how to choose the speeds in such a way that no permutation will be repeated until all the others have occurred.

Tins in a drawer

It is possible to pack tightly in a rectangular drawer, 186 mm by 286 mm, 16 equal cylindrical tins having the height of the drawer but such that any increase in their diameter will make it impossible to pack them all into the drawer. What is the diameter of the tin?

Bacilli

Dr. Abracadabrus discovered a type of rod-shaped bacteria which multiply in a strange way. From the original bacillus one part breaks off and becomes an independent bacillus; it is shorter than the remaining part, so that we have two individuals of different lengths. From the longer one a new bacillus, equal to the shorter one, breaks off; and the process continues until this breaking off leads to the residue of the original bacillus being shorter than the part broken off. Then from among the longest of the existing bacilli a part equal to the shortest of the existing bacilli breaks off. This one rule (together with unequal initial division) suffices to describe the whole multiplication process, but we must remember that at each moment at most one bacillus divides into two. Prove (1) that at no time are there more than three different lengths in the colony of bacilli, (2) that if the first division is irrational then there will sometimes be only two different lengths and sometimes three different lengths, and (3) that there is possible an initial division giving a ratio of lengths that will be preserved in all divisions.

The circus is coming

Children are playing in a meadow near the highway. Suddenly they see, at the place where the highway comes out of the forest, a clown riding a horse — there is a circus coming! The children wish to run to the highway in order to have a closer look at the clown. The children standing further away will not reach him, but they wish to see him from a distance as short as possible. All the children are running with equal speed, but the clown is riding faster.

The reader is asked to (a) draw the line separating the part of the meadow from which it is possible to reach the clown from the remaining part; (b) give the course of running for the children standing on this line; (c) give courses for those children who will not reach the highway before the clown passes them.

We advise the readers who attempt to solve this problem to become acquainted with problem 89.

Three cowboys

Three cowboys are looking after cattle in a large square pasture. The cowboys want to divide the pasture in such a way that firstly, each of them will be responsible for 1/3 of the area; secondly, the distance of each cowboy's post from the furthest part of his allotted share of the pasture will be the same. The cowboys would also like, thirdly, this maximum distance to be the smallest possible; and fourthly, they stipulate that in case of an accident the nearest of them should always ride to the spot in question.

Prove that this problem cannot be solved, and give the solution that can be obtained by the rejection of one or two conditions.

Investigation

The judge: So the witness did see the fire? What was the witness doing just before the fire?

The witness: I was walking between the fields.

The judge: How do the boundary lines of the fields run in your village?

The witness: Parallel and perpendicular to the highway.

The judge: Did the witness walk aimlessly?

The witness: No, going from the highway between the fields I looked at the fields of my neighbour. I never returned by the same way, and I reached the highway again.

The judge: Did the witness cross his own track?

The witness: No. But I remember that I walked by the field of barley twice, and the first time it was on my right side, so that I could see the hut, and when I walked the second time it was on my left. Then I heard a cry: Fire!

The judge ordered the witness to be arrested. Why?

Arrows on a dodecahedron

Let us imagine a model of Plato's dodecahedron. On each of its faces we shall draw an arrow →. Show that there will be found two neighbouring arrows, that is, arrows placed on adjoining faces, which contain an angle greater than 90°.

SOLUTIONS

1. We first of all remark that in a given sequence an odd digit can appear only between two even digits.

In fact, let us suppose that two consecutive terms c, d of the sequence considered are odd. Two cases occur:

either the number $\overline{cd}$ is the product of two terms a, b of the given sequence, which precede terms c, d and are odd,

or the digit c gives the product of two terms a, b of the sequence, which precede terms c, d and are odd.

In this way from the assumption that two consecutive terms c, d of the sequence in question are odd, it follows that two other terms of this sequence which precede terms c, d are odd. Thus from the assumption that any two consecutive terms c, d of the sequence are odd it would result that of three initial digits of the given sequence two are odd, which is not true.

From the observation that in the sequence considered no two neighbouring terms are odd it follows that the digit 9 never appears in the given sequence. In fact, the digit 9 could appear only in a product not less than 90, but the product of two one-digit numbers is always less than 90.

The digit 7 cannot appear either, as the only two-digit product of one-digit numbers one even and the other odd in which the digit 7 appears is the product $8 . 9 = 72$, and the digit 9 does not appear in the given sequence.

We shall not encounter in this sequence the digit 5 because only two products of one-digit numbers, one odd and the second even, contain the digit 5, namely $54 = 6 . 9$ and $56 = 7 . 8$, but they both contain digits which do not appear in our sequence.

2. Let us denote an arbitrary n-digit number by

$$L = 10^{n-1}a_n + 10^{n-2}a_{n-1} + \ldots + 10^2 a_3 + 10 a_2 + a_1$$

and the sum of the squares of its digits by

$$L_1 = a_n^2 + a_{n-1}^2 + \ldots + a_3^2 + a_2^2 + a_1^2.$$

We have

$$L - L_1 = (10^{n-1} - a_n)a_n + (10^{n-2} - a_{n-1})a_{n-1} + \ldots +$$
$$+ (10^3 - a_4)a_4 + (10^2 - a_3)a_3 +$$
$$+ (10 - a_2)a_2 - (a_1 - 1)a_1.$$

Let us remark that

$$(a_1 - 1)a_1 \leqq 72.$$

If we suppose that $n \geqq 3$, then (considering $a_n \neq 0$) we shall have

$$(10^{n-1} - a_n)a_n \geqq 99$$

and

$$(10^{i-1} - a_i)a_i \geqq 0 \quad \text{for} \quad i = 2, 3, \ldots, n-1,$$

and thus

$$L > L_1.$$

From the last inequality we find that, given a number L consisting of at least three digits and forming in the way indicated in the problem the sequence

$$L_1, L_2, L_3, \ldots \tag{1}$$

each number of the sequence being the sum of the squares of the digits of the preceding number, we shall obtain a sequence which decreases as long as its terms consist of at least three digits. Since these terms take only natural values, if we start from an arbitrary natural number L, which has not less than 3 digits, after a certain number of the operations described in the problem we must reach a number consisting of at most 3 digits. Hence we conclude that it suffices to find if the hypothesis stated in the problem is true for numbers having at most three digits.

Consequently let us suppose that we have a given number L consisting of 3 digits, that is $n = 3$. Thus $a_3 \neq 0$, and we have

$$L - L_1 = (100 - a_3)a_3 + (10 - a_2)a_2 - (a_1 - 1)a_1 \geqq 99 - 72 = 27;$$

thus

$$L_1 \leqq L - 27.$$

From this last inequality it follows that a certain term of sequence (1) is a number consisting of at most two digits. Let this number be

$$L_q = 10j+k.$$

The terms of the sequence

$$L_{q+1}, L_{q+2}, L_{q+3}, \ldots$$

are not changed when we replace the term L_q by $10k+j$; therefore it suffices to prove the assertion stated in this problem for numbers L_q under conditions

$$j \geqq k \geqq 0, \quad j \geqq 1.$$

When $L_q = 10j+k$, $j \geqq k \geqq 0$ and $j \geqq 1$ then L_{q+1} is one of the numbers of the following table of values of j^2+k^2:

j＼k	0	1	2	3	4	5	6	7	8	9
1	1	2								
2	4	5	8							
3	9	10	13	18						
4	16	17	20	25	32					
5	25	26	29	34	41	50				
6	36	37	40	45	52	61	72			
7	49	50	53	58	65	74	85	98		
8	64	65	68	73	80	89	100	113	128	
9	81	82	85	90	97	106	117	130	145	162

From this table we can reject the numbers

$$1, 10, 100$$

and the numbers

$$145, 20, 4, 16, 37, 58, 89,$$

mentioned in the problem, as this theorem is obvious for them. We can also reject in order the numbers

$$2, 40, 50, 52, 61, 73, 80, 81, 85, 90, 98, 130,$$

each of which is obtainable from an already rejected number, or from other numbers of the table, by either permuting the digits or ascribing 0.

Thus there will remain 28 numbers, namely

5, 8, 9, 13, 17, 18, 25, 26, 29, 32, 34, 36, 41, 45,

49, 53, 64, 65, 68, 72, 74, 82, 97, 106, 113, 117, 128, 162,

for which the validity of the theorem must be examined.

We present the results of this examination in a table. In the first column we shall write numbers for which we are verifying the theorem, and in the second column we shall write the consecutive terms of sequence (1) formed just for this number. We shall stop this examination when we reach one of the numbers for which the theorem is true:

5	25, 29, 85	72	53, 34, 25
8	64, 52	74	65
9	81	82	68, 100
18	65, 61	106	37
32	13, 10	113	11, 2
36	45, 41, 17, 50	128	69, 117, 51, 26, 40
49	97, 130	162	41

As in every case, we eventually reach the number 1 or one of the numbers

145, 42, 20, 4, 16, 37, 58, 89,

which appear periodically; the theorem stated in the problem is thus proved.

3. By dividing the powers 5^α, 4^β, 3^γ, by 11 (where α, β, γ denote non-negative integral numbers less than 5) we obtain the following remainders:

Number	Remainder	Number	Remainder	Number	Remainder
5^0	1	4^0	1	3^0	1
5^1	5	4^1	4	3^1	3
5^2	3	4^2	5	3^2	9
5^3	4	4^3	9	3^3	5
5^4	9	4^4	3	3^4	4
5^5	1	4^5	1	3^5	1

Let us denote respectively by $R(5^\alpha)$, $R(4^\beta)$, $R(3^\gamma)$ the remainders obtained by division of the numbers 5^α, 4^β, 3^γ by 11. These remainders can be read from the table given above. Let us denote by k, m, n three arbitrary non-negative integral numbers. The numbers 5^{5k}, 4^{5m}, 3^{5n} divided by 11 give the remainder 1. Thus numbers $5^{5k+\alpha}$, $4^{5m+\beta}$, $3^{5n+\gamma}$ divided by 11 give the corresponding remainders $R(5^\alpha)$, $R(4^\beta)$, $R(3^\gamma)$. In such a case the first expression

$$5^{5k+\alpha}+4^{5m+\beta}+3^{5n+\gamma} \tag{1}$$

is divisible by 11 when the sum $R(5^\alpha)+R(4^\beta)+R(3^\gamma)$ is divisible by 11. This occurs when $\alpha = 1$, $\beta = 2$, $\gamma = 0$.

It is possible to give 14 other expressions of type (1), divisible by 11. One finds all of them by choosing one number from each of the three columns of the above table, in such a way as to obtain a sum divisible by 11.

4. The expression a^n+b^n is divisible by $a+b$ if n is an odd number. Thus the number

$$3^{105}+4^{105} = (3^3)^{35}+(4^3)^{35}$$

is divisible by $3^3+4^3 = 7.13$. Similarly from the equations

$$3^{105}+4^{105} = (3^5)^{21}+(4^5)^{21},$$

$$3^{105}+4^{105} = (3^7)^{15}+(4^7)^{15}$$

it follows that it is divisible by $3^5+4^5 = 7.181$ and by $3^7+4^7 = 49.379$.

Let us remark that

$$4^3 \equiv -1 \;(\mathrm{mod}\; 5)$$

(this formula denotes that number 4^3 divided by 5 gives the remainder -1). It follows, therefore, that

$$4^{105} \equiv (-1)^{35} \;(\mathrm{mod}\; 5), \quad \text{thus} \quad 4^{105} \equiv -1 \;(\mathrm{mod}\; 5);$$

similarly

$$3^2 \equiv -1 \;(\mathrm{mod}\; 5),$$

whence

$$3^{104} \equiv (-1)^{52} \;(\mathrm{mod}\; 5), \quad \text{thus} \quad 3^{104} \equiv 1 \;(\mathrm{mod}\; 5),$$

and thus

$$3^{105} \equiv 3 \pmod 5;$$

now

$$4^{105} \equiv -1 \pmod 5 \quad \text{and} \quad 3^{105} \equiv 3 \pmod 5,$$

and consequently

$$3^{105} + 4^{105} \equiv 2 \pmod 5,$$

which means that the number $3^{105} + 4^{105}$ divided by 5 gives the remainder 2.

Similarly

$$4^3 \equiv -2 \pmod{11}, \quad \text{whence} \quad 4^{15} \equiv -32 \pmod{11},$$

and since

$$-32 \equiv 1 \pmod{11}, \quad \text{we get} \quad 4^{15} \equiv 1 \pmod{11},$$

and eventually

$$4^{105} \equiv 1 \pmod{11}.$$

In the same way we find that

$$3^5 \equiv 1 \pmod{11}, \quad \text{hence} \quad 3^{105} \equiv 1 \pmod{11}.$$

Thus

$$3^{105} + 4^{105} \equiv 2 \pmod{11},$$

which shows that the number $3^{105} + 4^{105}$ divided by 11 gives the remainder 2.

5. Let us suppose that there exist natural numbers x, y, z, n such that $n \geq z$ and $x^n + y^n = z^n$.

It is not difficult to see that $x < z$, $y < z$ and $x \neq y$; because of symmetry we can assume $x < y$.

Then

$$z^n - y^n = (z-y)(z^{n-1} + yz^{n-2} + \ldots + y^{n-1}) \geq 1 . nx^{n-1} > x^n,$$

contrary to the assumption that $x^n + y^n = z^n$. This contradiction implies the theorem asserted in the problem.

6. It is possible to give several sets of numbers x_1, x_2, ..., x_{10}, satisfying the conditions of the problem. Here are two of them:

0·95, 0·05, 0·34, 0·74, 0·58, 0·17, 0·45, 0·87, 0·26, 0·66;

0·06, 0·55, 0·77, 0·39, 0·96, 0·28, 0·64, 0·13, 0·88, 0·48.

Numbers of the first of these sets are distributed as follows in the closed interval $\langle 0, 1 \rangle$:

	2	3	4	5	6	7	8	9	10
0·95	2	3	4	5	6	7	8	9	10
0·05	1	1	1	1	1	1	1	1	1
0·34		2	2	2	3	3	3	4	4
0·74			3	4	5	6	6	7	8
0·58				3	4	5	5	6	6
0·17					2	2	2	2	2
0·45						4	4	5	5
0·87							7	8	9
0·26								3	3
0·66									7

This table is to be read as follows: In the column with the heading 2 we find number 2 in the first row and number 1 in the second row; this indicates that 0·05 lies in the first half of the interval $\langle 0, 1 \rangle$ and 0·95 in the second half of that interval. This rule applies to every column. Thus, for instance, number 6 in the column headed by 8 lies in the same row as 0·74; this indicates that 0·74 lies in the 6th subinterval of $\langle 0, 1 \rangle$ when we divide $\langle 0, 1 \rangle$ into 8 equal parts ($5/8 < 0·74 < 6/8$). We can verify all such statements summarized by the table. The numbers under any heading are all different, and as under every heading we have as many numbers as the heading indicates, the problem is solved.

7. There exists a set of 14 numbers ($n = 14$) satisfying conditions analogous to those given in the preceding problem:

0·06, 0·55, 0·77, 0·39, 0·96, 0·28, 0·64,

0·13, 0·88, 0·48, 0·19, 0·71, 0·35, 0·82

(it arises by enriching by 0·19, 0·71, 0·35, 0·82 the second set given in the former problem).

The distribution of the 14 numbers of this set in the interval $\langle 0, 1 \rangle$ is shown in the following table:

	2	3	4	5	6	7	8	9	10	11	12	13	14
0·06	1	1	1	1	1	1	1	1	1	1	1	1	1
0·55	2	2	3	3	4	4	5	5	6	7	7	8	8
0·77		3	4	4	5	6	7	7	8	9	10	11	11
0·39			2	2	3	3	4	4	4	5	5	6	6
0·96				5	6	7	8	9	10	11	12	13	14
0·28					2	2	3	3	3	4	4	4	4
0·64						5	6	6	7	8	8	9	9
0·13							2	2	2	2	2	2	2
0·88								8	9	10	11	12	13
0·48									5	6	6	7	7
0·19										3	3	3	3
0·71											9	10	10
0·35												5	5
0·82													12

Since numbers 0·35 and 0·39 are included between $5/15 = 0.33\ldots$ and $6/15 = 0.4$, it is impossible to complete the set given above with a 15th number without violating the conditions of the problem.

It is interesting that there exists a permutation of the above 14 numbers which preserves the conditions of the problem, for instance:

0·19, 0·96, 0·55, 0·39, 0·77, 0·06, 0·64,

0·28, 0·88, 0·48, 0·13, 0·71, 0·35, 0·82.

A. Schinzel has proved that the general solution is negative, by showing the unsolvability for $n = 75$ as follows. [1]

Let us suppose that the numbers $x_1, x_2, \ldots, x_{75}$ satisfy the required conditions. We then have for certain natural $i < j \leqq 35$

$$\frac{7}{35} < x_i < \frac{8}{35}, \qquad \frac{9}{35} < x_j < \frac{10}{35}. \tag{1}$$

[1] M. Warmus has proved quite recently the number $n = 17$ to be the last one for which the problem has a solution.

Hence

$$\frac{1}{x_j-x_i} + \frac{1}{x_i} < \frac{1}{\frac{9}{35} - \frac{8}{35}} + \frac{1}{\frac{7}{35}} = 40, \tag{2'}$$

$$\frac{x_j-x_i}{x_i} + x_j < \frac{\frac{10}{35} - \frac{7}{35}}{\frac{7}{35}} + \frac{10}{35} = \frac{5}{7}. \tag{2}$$

Let

$$k = \left[35(x_j-x_i) + \frac{5}{7}\right], \quad l = -\left[-\frac{(k+\frac{2}{7})x_i}{x_j-x_i}\right],$$

$$m = -\left[-\frac{l}{x_i}\right]$$

([x] denotes the integral part of x). We have the inequalities

$$35(x_j-x_i) - \frac{2}{7} < k \le 35(x_j-x_i) + \frac{5}{7}, \tag{3}$$

$$\frac{(k+\frac{2}{7})x_i}{x_j-x_i} \le l < \frac{(k+\frac{2}{7})x_i}{x_j-x_i} + 1, \tag{4}$$

$$\frac{l}{x_i} \le m < \frac{l}{x_i} + 1. \tag{5}$$

From (3) and (4) it follows that

$$35x_i < l < \frac{(35(x_j-x_i)+1)x_i}{x_j-x_i} + 1 = 35x_i + \frac{x_i}{x_j-x_i} + 1,$$

and considering (5) and (2') we get

$$35 < m < 35 + \frac{1}{x_j-x_i} + \frac{1}{x_i} + 1 < 76,$$

or

$$36 \le m \le 75. \tag{6}$$

Inequality (5) gives, moreover, $(m-1)x_i < l \le mx_i$, whence

$$[(m-1)x_i] < [mx_i]. \tag{7}$$

On the other hand, considering (5), (4) and (1)

$$(m-1)x_j \geqq \left(\frac{l}{x_i}-1\right)x_j = l\frac{x_j}{x_i} - x_j = l + l\frac{x_j-x_i}{x_i} - x$$

$$\geqq l + \frac{(k+\frac{2}{7})x_i}{x_j-x_i} \cdot \frac{x_j-x_i}{x_i} - x_j = l+k+\frac{2}{7} - x_j > l+k.$$

Similarly, considering (2)

$$mx_j < \left(\frac{l}{x_i}+1\right)x_j = l\frac{x_j}{x_i} + x_j = l + l\frac{x_j-x_i}{x_i} + x_j$$

$$< l + \left(\frac{(k+\frac{2}{7})x_i}{x_j-x_i}+1\right)\frac{x_j-x_i}{x_i} + x_j$$

$$= l+k+\frac{2}{7} + \frac{x_j-x_i}{x_i} + x_j < l+k+1.$$

Combining the above inequalities we obtain

$$l+k < (m-1)x_j < mx_j < l+k+1,$$

whence

$$[(m-1)x_j] = [mx_j]. \tag{8}$$

From (7) and (8) it follows that

$$N_{m-1} = [(m-1)x_j]-[(m-1)x_i] > [mx_j]-[mx_i] = N_m. \tag{9}$$

However, considering the conditions of the problem and inequality (6), the sequence $[(m-1)x_1]$, $[(m-1)x_2]$, ..., $[(m-1)x_{m-1}]$ is a permutation of the sequence 0, 1, ..., $m-2$, and similarly the sequence $[mx_1]$, $[mx_2]$, ..., $[mx_m]$ is a permutation of the sequence 0, 1, ..., $m-1$.

As $i, j \leqq 35 \leqq m-1$, N_{m-1} is the number of solutions of the inequality $x_i < x_t \leqq x_j$ for natural numbers $t \leqq m-1$, and N_m is the number of solutions of this inequality in natural numbers $t \leqq m$. Hence evidently $N_{m-1} \leqq N_m$, contrary to inequality (9). This contradiction completes the proof.

8. We can assume

$$A:B:C = \sqrt[3]{\frac{p}{r}} : \sqrt[3]{\frac{q}{p}} : \sqrt[3]{\frac{r}{q}}.$$

in fact, considering the obvious relation

$$pqr = 1,$$

we have

$$A:B = \sqrt[3]{\frac{p^2}{qr}} = p, \quad B:C = \sqrt[3]{\frac{q^2}{rp}} = q,$$

$$C:A = \sqrt[3]{\frac{r^2}{pq}} = r.$$

Using again the equation $pqr = 1$, we can obtain from the solution given above another one having the required properties:

$$A:B:C = \sqrt[3]{p^2q} : \sqrt[3]{q^2r} : \sqrt[3]{r^2p}.$$

9. For $x > 0$ the left side of the equation increases and it is easy to find that it is less than 10 for $x = 1\cdot5$ and greater than 10 for $x = 1\cdot6$. The root of the equation lies therefore in the open interval $(1\cdot5, 1\cdot6)$. Let us write it as an irreducible fraction p/q. The equation will take the form $p^5 + pq^4 = 10q^5$, from which it follows that p is a divisor of number 10, whence it is one of the numbers 1, 2, 5, 10. Now writing all the fractions p/q with $p = 1, 2, 5, 10$ we immediately see that none of them falls in the interval $(1\cdot5, 1\cdot6)$ whatever be the natural number q.

10. We first remark that we have the lemma: if p, q, x, y denote positive numbers, then the inequalities

$$\frac{1}{p} > \frac{1}{q} \quad \text{and} \quad x > y$$

lead to the inequality

$$\frac{x}{x+p} > \frac{y}{y+q}.$$

I n fact, from the assumption we have

$$\frac{1}{p} > \frac{1}{q} > 0 \quad \text{and} \quad x > y > 0,$$

and thus

$$\frac{x}{p} > \frac{y}{q} > 0, \quad \text{thus} \quad \frac{p}{x} < \frac{q}{y}.$$

If so, then

$$0 < 1 + \frac{p}{x} < 1 + \frac{q}{y}, \quad \text{or} \quad 0 < \frac{x+p}{x} < \frac{y+q}{y},$$

and thus

$$\frac{x}{x+p} > \frac{y}{y+q}.$$

The lemma is thus proved. Since A, B, C, a, b, c, r are positive numbers, then

$$\frac{1}{c+r} > \frac{1}{C+c+b+r} \quad \text{and} \quad A+a+B+b > A+a,$$

and thus according to the lemma,

$$\frac{A+a+B+b}{A+a+B+b+c+r} > \frac{A+a}{C+c+A+a+b+r}. \tag{1}$$

Similarly

$$\frac{1}{a+r} > \frac{1}{A+a+b+r} \quad \text{and} \quad B+b+C+c > C+c.$$

Thus, by the lemma,

$$\frac{B+b+C+c}{B+b+C+c+a+r} > \frac{C+c}{C+c+A+a+b+r}. \tag{2}$$

Adding the right and left sides of inequalities (1) and (2) respectively, we obtain the inequality the validity of which was to be established.

11. The equation $1-z = z^2$ has one and only one positive root $z = (\sqrt{5}-1)/2$. Hence this z satisfies the equations

$$1-z = z^2, \quad z-z^2 = z^3, \quad \ldots, \quad z^n-z^{n+1} = z^{n+2}, \ldots$$

Then assigning to the elements a_n the values z^n ($n = 0, 1, 2, \ldots$) we obtain the required sequence. As $0 < z < 1$, the limit of this geometrical progression is zero. Now let b_0, b_1, ... be another sequence satisfying the conditions of the problem. We must have $b_0 = 1$ and $b_n-b_{n+1} = b_{n+2}$ for $n = 0, 1, \ldots$ If this sequence is not identical with the sequence found before, then there is a certain element b_k differing from z^k, and thus $b_k = z^k+d$, and d is not equal to zero. Let k be the first subscript for which this difference appears; we know that $a_0 = 1 = z^0$, and thus k is at least 1, so that there exists an element b_{k-1} for which $b_{k-1} = z^{k-1}$. Thus we have

$$
\begin{aligned}
b_k &= z^k+d, \\
b_{k+1} &= b_{k-1}-b_k = z^{k-1}-(z^k+d) = z^{k-1}-z^k-d = z^{k+1}-d, \\
b_{k+2} &= b_k-b_{k+1} = z^k+d-(z^{k+1}-d) = z^k-z^{k+1}+2d \\
&= z^{k+2}+2d, \\
b_{k+3} &= b_{k+1}-b_{k+2} = z^{k+1}-d-(z^{k+2}+2d) = z^{k+3}-3d, \\
b_{k+4} &= z^{k+3}+5d, \\
b_{k+5} &= z^{k+4}-8d, \\
&\cdots \cdots \cdots \cdots \\
b_{k+p} &= z^{k+p} \pm C_p d.
\end{aligned}
$$

Since z^{k+p} tends to zero as $p \to \infty$, then starting from a sufficiently large p it is the second term $\pm C_p d$ which determines the sign of the term b_{k+p}, the absolute value of $C_p d$ being always not less than d. But the integral coefficients $\pm C_p$ multiplying d are positive and negative alternatively, and thus in the sequence $\{b_n\}$ there must appear a negative element, which contradicts what has been required. Thus the assumption of non-identity leads to a contradiction, and consequently the sequence $\{z^n\}$ is the unique solution.

12. I. Let us suppose that the resulting figure includes a closed polygon *ABCDE...MN*.

Moreover, let us suppose that $AN < AB$, that is, the point *A* is connected with the point *N* as the nearest one. Then $AB < BC$. But the points *B* and *C* are connected by a segment, and thus $BC < CD$. Continuing this line of reasoning, we obtain $CD < DE < ... < MN < NA$, i.e. $AB < NA$, which contradicts the assumption $AN < AB$.

Under the assumption $AN > AB$ the same argument can be employed, and thus this assumption also leads to a contradiction. Hence, it follows that the newly formed figure cannot include a closed polygon.

II. Let us suppose that the newly formed figure includes two intersecting segments, *AB* and *CD* (Fig. 4).

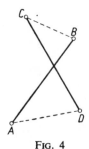

Fig. 4

Let us suppose that the points *A* and *B* have been connected by a segment because the point *B* lies nearest to the point *A*; and let us suppose similarly that the point *D* lies nearest to the point *C*. Then $AB < AD$, $CD < CB$, whence $AB + CD < AD + CB$, which contradicts the proposition that in a convex quadrangle the sum of the diagonals is greater than the sum of two opposite sides.

In this way we have proved the second part of the theorem.

13. Let us assume for simplicity of notation

$$r = \sqrt{x_1^2 + \ldots + x_{k-1}^2}, \quad a = x_k, \quad b = x_{k+1}.$$

We shall show that if $a < b$, then the order

$$x_1, x_2, \ldots, x_{k-1}, b, a, x_{k+2}, \ldots, x_n$$

gives a smaller angle than the original order

$$x_1, x_2, \ldots, x_{k-1}, a, b, x_{k+2}, \ldots, x_n.$$

For this purpose it suffices to compare the angle $P_{k-1}OP_{k+1}$ in the two orders. In Fig. 5 the angle $P'_{k+1}OP_{k-1}$ corresponds to the

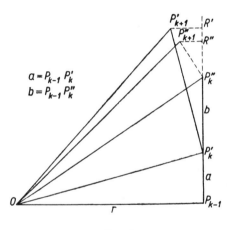

FIG. 5

original order, and the angle $P''_{k+1}OP_{k-1}$ corresponds to the changed order. As $OP'_{k+1} = OP''_{k+1} = \sqrt{r^2 + a^2 + b^2}$, we must prove the inequality $P''_{k+1}R'' < P'_{k+1}R'$, which is

$$\frac{ab}{\sqrt{b^2 + r^2}} < \frac{ab}{\sqrt{a^2 + r^2}}$$

and follows from the assumption $a < b$.

14. If in the triangle ABC with sides a, b, c there is given an angle A equal to $60°$, then angle B + angle $C = 120°$, and six such triangles can be formed into a garland (Fig. 6) limited from the outside by a regular hexagon with side a, and limited from the inside by a regular hexagon with side $b-c$. Computing the area of both hexagons we obtain formula (1), given in the problem.

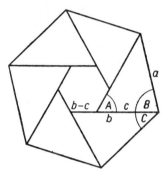

FIG. 6

In the case where angle A in the triangle is $120°$ we have angle $B+$angle $C = 60°$ and three such triangles make up a triangular garland (Fig. 7). Computations similar to the former one give formula (2).

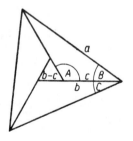

FIG. 7

15. Let us draw a straight line which intersects sides $A_1 A_2$ and $A_3 A_1$ of the triangle $A_1 A_2 A_3$ and halves its perimeter (equal to $2p$) (Fig. 8). Let us take straight lines $A_1 A_2$ and $A_1 A_3$ as the co-ordinate axes and let the straight line $C_3 C_2$ pass through two points $C_3(q+\lambda, 0)$ and $C_2(0, q-\lambda)$, where $q = p/2$, and λ denotes an arbitrary numbei. The straight line $C_3 C_2$ obviously halves the perimeter of the triangle $A_1 A_2 A_3$.

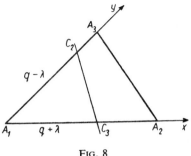

FIG. 8

From the condition that the straight line $C_2 C_3$ intersects sides $A_1 A_2$ and $A_3 A_1$ but not their extensions, we have $0 \leqq q+\lambda \leqq d_3$, $0 \leqq q-\lambda \leqq d_2$ (d_1, d_2, d_3, are the lengths of $A_2 A_3$, $A_3 A_1$, $A_1 A_2$ respectively); hence

$$-q \leqq \lambda \leqq d_3 - q, \qquad q - d_2 \leqq \lambda \leqq q. \tag{1}$$

It is evident that from the inequality $d_1 < d_2 + d_3$ we have $2d_1 < 2p$, i.e. $d_1 < p$, and similarly $d_2 < p$, $d_3 < p$. It is also clear that from the inequality $p > d_3$ results the inequality $q > d_3 - q$, and that, similarly, from the inequality $p > d_2$ we have the inequality $q - d_2 > -q$. Thus, considering this, the system of inequalities (1) can be replaced by a double inequality

$$q - d_2 \leqq \lambda \leqq d_3 - q, \tag{2}$$

giving a bound for the parameter λ.

The equation of the straight line $C_3 C_2$ is

$$(q - \lambda)x + (q + \lambda)y - q^2 + \lambda^2 = 0. \tag{3}$$

If, having drawn the triangle $A_1 A_2 A_3$ we measured its perimeter $2p$, and then, taking successively different numerical values of the parameter λ, we found for each value of λ the points C_3,

C_2 and drew a straight line $C_3 C_2$, then we would realize (Fig. 9)
that all straight lines $C_3 C_2$ are tangent to the same curve, which is
the envelope of the straight lines (3).

By computation it is easy to find the equation of this curve as

$$4xy = (p-x-y)^2. \qquad (4)$$

Curve (4) is a parabola, because it is a curve of the second degree
which has only one common point with any straight line defined
by the equation $y = x+k$, independently of the value of the

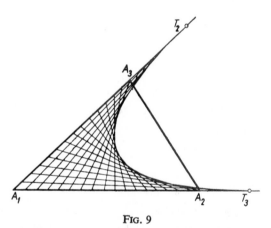

FIG. 9

parameter k. The axis of parabola (4) is thus parallel to the straight
line $y = x$, i.e. to the bisector of the angle A_1. Putting in equation
(4) successively the values $y = 0$ and $x = 0$, we can prove that
parabola (4) is tangent to the arms of the angle A_1 at points
$T_3(p, 0)$ and $T_2(0, p)$ respectively, which is shown in Fig. 10.
And as $A_1 T_2 T_3$ is an isosceles triangle, the bisector of angle A_1
is perpendicular to $T_2 T_3$ and halves this segment, which shows
the bisector to be the axis of parabola (4). Now taking successively
the extreme values of the parameter λ, defined by inequality (2),
we obtain for $\lambda = q-d_2$ a straight line $A_3 B_3$ tangent to parabola
(4) at the point P_3 whose coordinates are

$$x = \frac{(p-d_2)^2}{p}, \qquad y = \frac{d_2^2}{p}$$

and for $\lambda = d_3 - q$ we obtain a straight line $A_2 B_2$ tangent to parabola (4) at the point P_2 whose coordinates are

$$x = \frac{d_3^2}{p}, \quad y = \frac{(p-d_3)^2}{p},$$

where $A_1 B_3 = p - d_2$, $B_3 A_2 = p - d_1$. The collinearity of the points T_2, B_1, P_2 and that of the points T_3, B_1, P_3 is easy to verify when B_1 denotes the point dividing the side $A_2 A_3$ into segments $p - d_3$ and $p - d_2$. It is also evident that the straight lines $T_2 B$

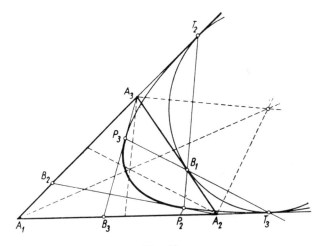

FIG. 10

and $T_3 B_1$ are respectively parallel to the bisectors of the angles A_3 and A_2.

Finally, it can be seen that points B_1, T_2, T_3 are points of tangency of the escribed circle.

Therefore any straight line halving the perimeter of a triangle $A_1 A_2 A_3$ and intersecting sides $A_1 A_2$ and $A_3 A_1$ of the triangle, is tangent to the arc of parabola (4) and *vice versa*.

The same reasoning can be applied to any of the three angles of the triangle. In this way we obtain three parabolic arcs, which, together with the segments of the common tangents, give rise to a curvilinear triangle T (Fig. 11).

Now we can formulate the following consequences:

(1) Through every point P, lying inside the triangle $A_1 A_2 A_3$ but outside the triangle T, there passes one and only one straight line halving the perimeter of the given triangle. It is a tangent which can be drawn from the given point to one of three arcs of the parabolas shown in Fig. 11.

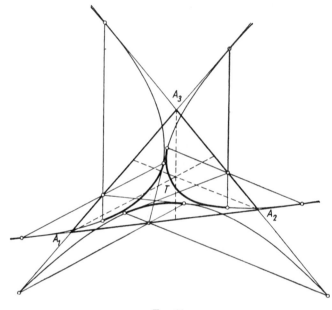

FIG. 11

(2) Through every point P lying inside of the triangle T there pass 3 and only 3 straight lines halving the perimeter of the given triangle $A_1 A_2 A_3$. There are three tangents which can be drawn from the point P to the arcs of the parabolas shown in Fig. 11.

(3) If we are given a direction in the plane, we can always draw a straight line having this direction and halving the perimeter of the triangle. For this purpose it is sufficient to draw a line tangent to one of the arcs of the parabola and having the given direction.

(4) If we are given two directions, we can always draw two straight lines having those directions and such that each of them halves the perimeter of the triangle. The common point Q of these lines

lies inside the curvilinear triangle T; considering this, if we are given two straight lines each halving the perimeter of a given triangle, we can always draw through their point of intersection, Q, a third straight line halving the perimeter of the triangle.

(5) If we are given a system of three straight lines l_1, l_2, l_3 (Fig. 12), rigidly connected (for instance drawn with India ink on Cellophane), we can always put this system on a plane so that each of these lines l_1, l_2, l_3 will halve the perimeter of the triangle $A_1 A_2 A_3$.

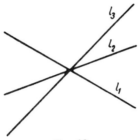

FIG. 12

The solution of this question has given a rather unexpected result. There is no need to look for the point having the property required in the problem as every point called Q in the text of the problem has it already.

If we replaced the triangle by another figure, for instance by a quadrangle, the solution of the problem would be similar. Instead of the curvilinear triangle T there would appear a space bounded by four arcs of parabolas, however, we could still, as in the triangle, draw one of the 3 straight lines halving the perimeter through any given point lying inside the quadrangle. In the case of a figure with a centre of symmetry, the domain through which one can lead three straight lines halving the perimeter reduces to a point, and any straight line passing through this point halves the perimeter.

16. To obtain a division of a triangle into a net of triangles such that at every vertex of the newly formed figure there meet the same number of sides, we employ classic polyhedra with triangular faces. These are the tetrahedron, the octahedron and the icosahedron, and these are the only ones.

If, outside of the tetrahedron, we choose one point lying near to the centre of one of the faces and project from this point the edges of the tetrahedron upon a plane, then we shall obtain the figure shown in Fig. 13. It consists of three triangles corresponding to

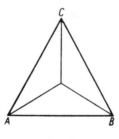

FIG. 13

three faces of the tetrahedron. The fourth face is projected onto the large triangle *ABC*. At every vertex of the figure there meet three faces, just as at every vertex of the tetrahedron there meet three edges.

FIG. 14

FIG. 15

In a similar way, from the regular octahedron we obtain by central projection a figure consisting of seven triangles in which every vertex belongs to 4 faces (Fig. 14), and from the regular icosahedron a net consisting of 19 triangles in which every vertex belongs to 5 faces (Fig. 15).

No net other than those shown in Figs. 13–15 satisfies the condition of the problem, because to such a net there would correspond a regular polyhedron different from the three mentioned above, and such a polyhedron does not exist. [Cf. *Mathematical Snapshots* (Oxford Univ. Press, 1960), the Chapter on the platonic solids; there is a proof that regular nets with triangular meshes ($f = 3$) are only such as are given by (a), (b), and (c); the respective values of F (= number of faces) are 4, 8 and 20.]

17. The answer to the question is in the affirmative. As there is only a finite number of straight lines passing through two of the given $3n$ points, we can choose a direction such that no straight line with this direction does satisfy this condition. When we move such a straight line on the plane without changing its direction and start far from all the points, at first all the points will lie on one side of the straight line, then it will touch successively the first, the second, the third, ..., and $3n$-th point. The consecutive positions of the line after it has left behind 3, 6, ..., $3n-3$ points divide the plane into strips, and in each of these strips there is only one triangle.

In the same way we can form quadrangles, pentagons etc. nonintersecting and not contained in one another.

18. The answer to the question in the problem is in the affirmative. It is possible to attach to the nodes of the network the signs

FIG. 16   FIG. 17

plus and minus in such a way as to satisfy the conditions of the problem.

Let us begin with the remark that the solution must satisfy one of the following conditions:

(1) At the three vertices of every equilateral triangle there are only plus signs (Fig. 16).

(2) At two vertices of every triangle there are minus signs and at one vertex there is a plus sign (Fig. 17).

We exclude the first case.

Connecting the triangles of Fig. 17 by the vertices of equal signs, we can form an infinite strip (Fig. 18).

We then see on the edge of the strip there appears a sequence of signs composed of triads $+ - -$. Hence by adjoining properly the identical strips of Fig. 18 we can cover the whole

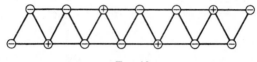

FIG. 18

plane with a diagram of equilateral triangles satisfying the conditions of the problem (Fig. 19).

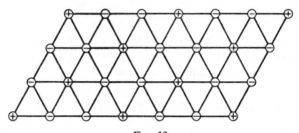

FIG. 19

From the construction of this diagram, satisfying the conditions of the problem, it follows that the location of plus and minus signs at the nodes is unique.

19. Let us assume that the contrary holds. That is, assume that it is possible to cover the plane in such a way that, at every node

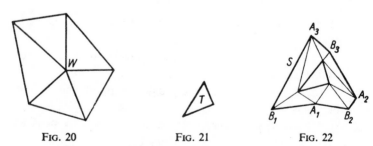

FIG. 20   FIG. 21   FIG. 22

W, 5 triangles are joined (Fig. 20). Then the sum of any four angles from among the five at the vertex W would have to be greater than $180°$.

Let us take an arbitrary triangle T belonging to the diagram (Fig. 21). The triangles of this diagram which have at least one common point with the triangle T would form a hexagon S (Fig. 22); at each of its vertices A_1, A_2, A_3 there would meet three triangles, and at each of the remaining three vertices B_1, B_2, B_3 there would meet two triangles. The hexagon S would have a pair of triangles adhering to it at each of the vertices A_1, A_2, A_3 (Figs. 23, 24) and one more triangle, in addition to the above-mentioned pairs, adhering to it at each of the vertices B_1, B_2, B_3 (Fig. 24). The whole

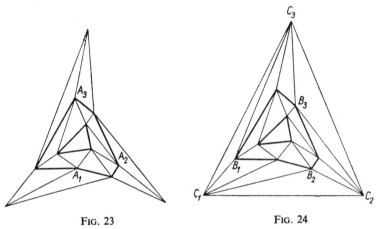

FIG. 23   FIG. 24

figure consisting of hexagon S and strips of adhering triangles would form the triangle $C_1 C_2 C_3$ (Fig. 24). At every vertex of this triangle there would meet four triangles of the diagram and the sum of four angles at the vertex would be less than 180°, which contradicts the assumption made earlier.

This contradiction leads to the statement that the diagram described in the problem does not exist.

20. Let $OA_0 = a$, $OB_0 = b$ (Fig. 25). According to the condition of the problem, $b = aq$, where

$$q = \tfrac{1}{2}(\sqrt{5}-1).$$

The side of the first of the squares cut out is aq, of the second aq^2, of the third aq^3 etc.

Let us put $OA_n = x_n$, $A_nB_n = y_n$ $(n = 1, 2, \ldots)$. Segments $OA_1 = x_1$, $OA_3 = x_3$, $OA_5 = x_5$, ... form an increasing sequence, and segments $OA_2 = x_2$, $OA_4 = x_4$, $OA_6 = x_6$, ... form a decreasing sequence. Both sequences approach the same limit, which is the abscissa x of a certain point A. Since $OA_1 = aq$, $A_1A_3 = aq^5$, $A_3A_5 = aq^9, \ldots$, we have

$$x = \lim x_{2n+1} = aq + aq^5 + aq^9 + \ldots = \frac{aq}{1-q^4}.$$

Similarly, the segments $y_1 = A_1B_1$, $y_3 = A_3B_3$, $y_5 = A_5B_5$, ... form a decreasing sequence, and the segments $y_2 = A_2B_2$,

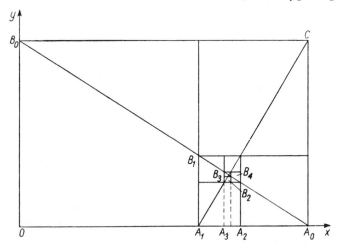

FIG. 25

$y_4 = A_4B_4$, $y_6 = A_6B_6, \ldots$ form an increasing sequence. Both sequences approach the same limit, this limit being the coordinate y of the point A. We have

$$y = \lim y_{2n} = aq^4 + aq^8 + aq^{12} + \ldots = \frac{aq^4}{1-q^4}.$$

From the rectangle only the point A will remain; its coordinates are

$$x = \frac{aq}{1-q^4}, \quad y = \frac{aq^4}{1-q^4}.$$

The reader will verify that the point A lies on the intersection of two perpendicular straight lines $A_0 B_0$ and $A_1 C$.

21. Let us draw a square S of side 1. If two opposite corners both belong to the same part, then the division evidently shares the property asserted, as the diagonal of S is $\sqrt{2}$, which is more than $\sqrt{65/64} > 1\cdot0077$. If no pair of opposite vertices belongs to the same part, we have to consider two cases: (1) there is a part containing no vertices of S, (2) each part contains one or two vertices. Fig. 26 symbolizes case (1). Let us study it.

We denote the part containing vertices A', A'' by A, the part containing vertices B', B'' by B, and the part containing no vertices by C. The vertical segments $A'N$, $A''M$ have the lengths 1/4 and 1/2, respectively. If M belongs to part A, then we take $P = M$, $Q = A'$ and get $PQ = \sqrt{5/4} > \sqrt{65/64}$, a confirmation of the property announced. If M belongs to B, then we put $P = M$, $Q = B'$ and get $PQ = \sqrt{5/4} > \sqrt{65/64}$ as before. If N belongs to A we

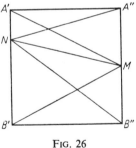

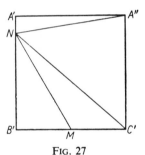

FIG. 26   FIG. 27

put $P = N$, $Q = A''$ with the result $PQ = \sqrt{17/16} > \sqrt{65/64}$; if it belongs to B we put $P = N$, $Q = B''$ and get $PQ = 5/4 > \sqrt{65/64}$. If neither M nor N belongs to the parts A, B, both must necessarily belong to C; then we put $P = M$, $Q = N$, which gives $PQ = \sqrt{17/16} > \sqrt{65/64}$.

Figure 27 symbolizes case (2); let us study it.

We denote the part containing vertices A', A'' by A, the part containing vertex B' by B, and the part containing vertex C' by C. The midpoint M of the base of S can belong to A, to B or to C.

If M belongs to A we put $P = M$, $Q = A'$ and get $PQ = \sqrt{5/4}$. If M belongs to B let us consider point N lying on the left side of S at a distance $1/8$ below A'. We have to study three possibilities: (i) N belongs to A; (ii) N belongs to B; (iii) N belongs to C.

We put $P = N$ whatever part N belongs to, but

we choose $Q = A''$, if (i), and get $PQ = \sqrt{65/64}$;

we choose $Q = M$, if (ii), and get $PQ = \sqrt{65/64}$;

we choose $Q = C'$, if (iii), and get $PQ = \sqrt{113/64} > \sqrt{65/64}$.

If, finally, M belongs to C we consider a point N on the right side of S, at a distance $1/8$ below the right upper corner A'', and proceed as in the case where M belongs to B; the oblique lines of Fig. 27 will be replaced by other such lines, meeting at the new point N, and the argument will be the same as previously, the rôles of B and C being interchanged. Thus the proof is finished.

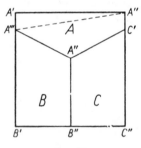

FIG. 28

Let us examine the question whether we can improve the theorem by replacing the number $\sqrt{65/64}$ by a greater number (for instance 1·01). The answer is negative. It is to be found in the division of the unit square into three parts A, B, C, of equal areas, shown in Fig. 28. The trapezoids B and C are congruent and their common vertical side has a length of $11/24$. The lengths of the vertical sides of the pentagon are $1/8$ each. The sketch is to be read as follows:

The interior of the pentagon belongs to A, the interior of the left trapezoid to B, that of the right trapezoid to C. The sides of pentagon belong to A, with exception of one point marked C'.

The sides of the left trapezoid belong to B with exception of the upper side $A'''A^{IV}$ belonging to A. The sides of the right trapezoid belong to C with the exception of the short vertical side A^{IV} B (of length 11/24).

It is easily seen that the longest diagonals of the polygons considered are equal to $\sqrt{65/64}$. It follows that it is impossible to find a pair of points P, Q belonging both to the same part if the distance PQ surpasses $\sqrt{65/64}$. The dotted segment $A'''A''$ has a length equal to $\sqrt{65/64}$; its extremities both belong to A. We note that this pair of points is the only one having the property stated in the text of the problem.

22. Labelling of the nodes of the lattice in conformity with the conditions of the problem is impossible.

To prove it, let us assume that it is possible to denote the nodes of the network in the way the conditions of the problem demand

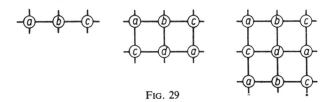

Fig. 29

and let us consider an arbitrary row of the diagram. In this row we must find a place with three different letters in three successive nodes, for example a, b, c (otherwise the row would contain at most two different letters, which contradicts the assumption).

In the next row (the middle part of Fig. 29) below the nodes denoted by a, b, c we must find corresponding letters c, d, a, if the labelling of nodes is to satisfy the conditions of the problem. This implies the marking of the nodes of the next row with corresponding letters a, b, c (right part of Fig. 29).

Continuing this line of reasoning we see that in each three columns of this part of the diagram there appear only two different letters: in the first there are only letters a, c, in the second b, d, and in the third c, a.

Thus it is impossible to find four different letters at the nodes of any row or column and we cannot mark the diagram in the way indicated in the problem.

23. If there were two points (x, y) and (u, v) on the circle with the centre $(\sqrt{2}, \sqrt{3})$ then we should have the relation

$$(x-\sqrt{2})^2+(y-\sqrt{3})^2 = (u-\sqrt{2})^2+(v-\sqrt{3})^2,$$

and hence

$$c\sqrt{2}+d\sqrt{3} = u^2+v^2-x^2-y^2 = n$$

where $c = 2(u-x)$, $d = 2(v-y)$ and n are integral numbers. Hence it follows that

$$2c^2+3d^2+2cd\sqrt{6} = n^2.$$

As $\sqrt{6}$ is an irrational number, and c, d and n integral numbers, then for $cd \neq 0$ we obtain a contradiction. Thus we should have $cd = 0$. If $c = 0$ then $d\sqrt{3} = n$ and we must have $d = 0$, and $n = 0$; similarly, $d = 0$ gives $c = 0$. Thus we must have $c = d = 0$, and from this it immediately follows that $x = u$, $y = v$; hence the points are identical.

24. The definition of the lattice of integral numbers: cf. problem 22.

From the solution of problem 23 it follows that a circle with centre at the point $(\sqrt{2}, \sqrt{3})$ whose radius is r contains $f(r)$ points of the lattice and $f(r)$ has the property of increasing by unit jumps as r increases. We shall show that for small r, $f(r)$ is zero and for large r $f(r)$ takes arbitrary large values. It is evident that for $r = 0\cdot1$ we have $f(r) = 0$. Every square with sides parallel to the x, y-axes and greater than natural n contain at least n^2 of the lattice points, because between the straight lines $x = a$ and $x = a+n$ there are at least n perpendicular rows of the lattice, and between the straight lines $y = b$ and $y = b+n$ there are at least n horizontal rows of the lattice. But a circle with radius greater than n contains a square with sides greater than n (and parallel to the axes), and thus it contains at least n^2 lattice points, whence $f(r)$ is greater than n^2 for $r = n+1$; and this is the only thing required in the proof, whose essence is that an unbounded function increasing from zero by unit jumps takes all positive integral values.

25. The error lies in the fact that the average of the terms of an infinite sequence depends on the permutation of these terms (for example, the average of the terms 1, 0, 1, 0, 1, 0, ... is the number $1/2$ and the average of terms of the same sequence but permutated as follows

$$1, 0, 1, 0, 0, 0, 0, 1, 0, 0, 0, 0, 0, 0, 0, 0, 0, 1, ...$$

is zero). The alleged equation $14 = 15$ results precisely from the different ways of arranging an infinite sequence whose terms are angles of heptagons covering the plane.

26. Let us assume that for n points, of which no three lie on the same straight line, we can always find closed polygons with non-intersecting sides whose vertices are those points. And let us choose $n+1$ points in such a way that no three among them lie on the

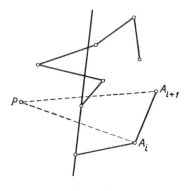

Fig. 30

same straight line. Among those points there exists one point, for example point P, which can be separated by a straight line from other points. Let W_n be a polygon which, according to the given assumption, satisfies the conditions of the problem, and whose vertices are points separated from the point P.

The answer to the question put in this problem will be in the affirmative if we can show that at least one of the sides of the polygon W_n is entirely visible from the point P, because replacing this side by two segments (Fig. 30) connecting point P with the ends of this side, we obtain a polygon W_{n+1} satisfying the conditions of the problem. Let us choose any side of the polygon W_n, for example the side $A_i A_{i+1}$. If this side is not entirely visible from the point P, i.e. if it is covered entirely or partially by the polygon W_n, then we extend this side and reject all those sides which are separated entirely from the point P by the newly formed straight line. After such an operation the number of sides not rejected decreases by at least one (namely the side $A_i A_{i+1}$), and we repeat this operation with respect to the remaining sides, choosing again an arbitrary side.

After at most n repetitions of this operation, nothing will remain on the side of the point P; on the last straight line there lies a side entirely visible from the point P. Removing this side and joining its ends with the P we obtain the desired $(n+1)$-gon.

Since the hypothesis stated in the problem is evident in the case of $n = 3$, the affirmative answer to the question put in the problem can be considered as proved by induction.

27. We draw a circle through the points *1*, *2*, *3* and a circle through the points *1*, *2*, *4*. If the point *4* lies in the circle *123*, or the point *3* lies in the circle *124*, then the answer to the question is affirmative. Let us assume that neither of these possibilities occurs. It is easy

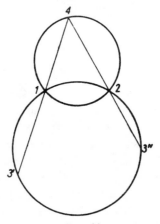

FIG. 31

to see that an arc of the circle *123* lying outside the circle *124*
is divided into three parts, *13'*, *3'3"* and *3"2* (Fig. *31*), such that
if the point *3* lies on *13'* then the point *1* lies in the circle *234*,
if *3* lies on *3"2*, then *2* lies in *134*, and if *3* lies on *3'3"* then both
cases occur.

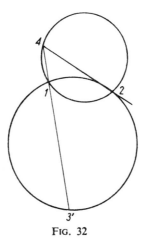

Fig. 32

We make the additional remark that one (Fig. 32) or two of the
above-mentioned three parts of the arc *12* can disappear. Never-
theless, in every case the answer to the question put in the problem
is in the affirmative.

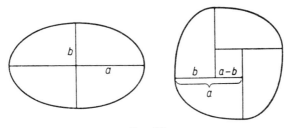

Fig. 33

28. We obtain the desired curve by cutting the ellipse into four
parts and rearranging them as in Fig. 33.

But there is another solution which does not demand removing one part of the ellipse from the plane: let us connect the suc-

cessive vertices of the ellipse by chords (Fig. 34). We obtain a rhombus lined with four segments of the ellipse. Let us replace this rhombus, whose area is $2ab$, by a square with area $c^2 = a^2+b^2$, so that the four segments of the ellipse still adhere to the sides

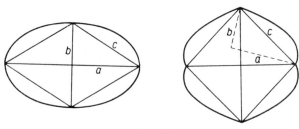

FIG. 34

of the square. The area enclosed by this curve will increase in such a way that the difference between the areas of the square and the rhombus is $c^2-2ab = a^2+b^2-2ab = (a-b)^2$, as in the former solution.

29. Instead of dividing the whole space, we divide a sphere, through whose centre we pass planes. On the surface of the sphere mutually intersecting orthodromes are formed. We consider one of them to be the equator, and we project all orthodromes from the centre of the sphere on a plane tangent to the sphere at the pole. The projections of orthodromes (with the exception of the one that is the equator) are straight lines. We have to compute the maximum number of domains of a plane divided by $n-1$ straight lines. By induction we can easily see that it is equal to $1+1+2+\ldots+(n-1)$ $= 1+\frac{1}{2}n(n-1)$, because considering $k-1$ existing straight lines, the k-th line can increase the number of domains by at most k. As on the sphere there are twice as many domains as in its plane projection, the number we seek is twice as large as the formerly computed number. Hence it is equal to $n(n-1)+2$. In particular, for $n = 4$ the number is 14.

30. It is easy to realise that the transformation we refer to in the problem is an inversion. In fact let us consider the section of the

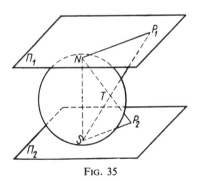

Fɪɢ. 35

globe and of both planes tangent to it by a plane passed through the axis NS (Figs. 35, 36). Let r_1 and r_2 be the distances of the points

P_1 and P_2 (corresponding to each other by the transformation defined) to N and S respectively. If we denote the radius of the sphere by r, then from the similarity of the right triangles NTP_1, STN, P_2TS we have the relation $r_1 r_2 = 4r^2$, defining an inversion.

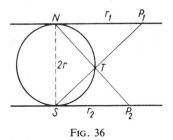

FIG. 36

31. The yarn wound on the cube rotating about one of its axes (Fig. 37) will remain only on the edges which have no common points with the axis of rotation. The yarn will cover half of each side of the cube, that is, half of the area of the surface of the cube.

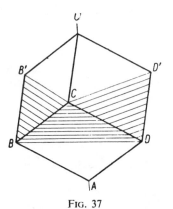

FIG. 37

Let us now rotate the cube successively about each of the four axes, but each time winding yarn of a different colour.

For rotation about axis AC' we shall use black yarn (say a), for rotation about DB' red yarn (say b), for rotation about BD' yellow yarn (say c), and lastly, for rotation about CA' let us take blue yarn (say d).

Then the cube will be covered in a way shown in Fig. 38; each face is divided into four triangles; the letters in each triangle denote the colours of the yarn covering this triangle.

It is easy to see that:

(1) On the surface of the cube there will appear six shades, i.e. as many as there are combinations of 4 elements taken 2 at a time, namely *ab*, *ac*, *ad*, *bc*, *bd*, and *cd*.

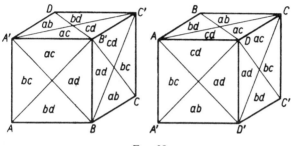

FIG. 38

(2) On every face there will be four different shades.

(3) The surface of the cube will be covered with two layers of yarn.

(4) Opposite faces of the cube will be coloured with the same shades but the cyclic order of their arrangement will be reversed.

32. We shall show that through every point on the surface of the cube there pass four different geodesics; in general we have seven families of geodesics.

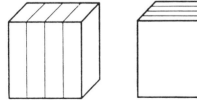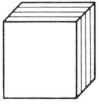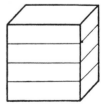

FIG. 39

If we assume that the cube is smooth, then the rubber band put on the cube will place itself in such a way that the perimeter of the polygon formed by it reaches its minimum.

Three kinds of such geodesics are given in Fig. 39; they lie on planes parallel to the faces of the cube. In order to ascertain that there are other possible families of geodesics, let us cut the

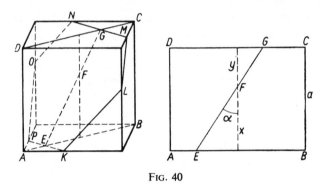

FIG. 40

cube with a plane parallel to the diagonal of the base (Fig. 40). Then following the notation used in Fig. 40 we have

$$x+y = a,$$

$$PK = a\sqrt{2}-2x\tan\alpha, \qquad KL = x\sqrt{1+2\tan^2\alpha},$$

$$MN = a\sqrt{2}-2y\tan\alpha, \qquad LM = y\sqrt{1+2\tan^2\alpha},$$

and for the perimeter p of the hexagon $KLMNOP$ we obtain

$$p = 2a\sqrt{2}-2a\tan\alpha+2a\sqrt{1+2\tan^2\alpha}.$$

Therefore the perimeter p depends only on the angle α and it remains the same in all parallel planes; this perimeter attains its minimum for $\tan\alpha = 1/\sqrt{2}$. The sides of the hexagon $KLMNOP$ are then parallel to the diagonal of the faces of the cube, and this hexagon is a geodesic. We have four families of such geodesics, as shown in Fig. 41; considering the former three we get altogether seven families.

33. Let us look at the system of 6 cubes shown in Fig. 42. We find that the cube B is the image of cube A by reflection with respect to their common face. In the same way C is derived from B, D from C, E from D, and F from E. Instead of consider-i ng the movement of the molecule in the cube A according to the l aw of reflection, we can observe the movement of a molecule

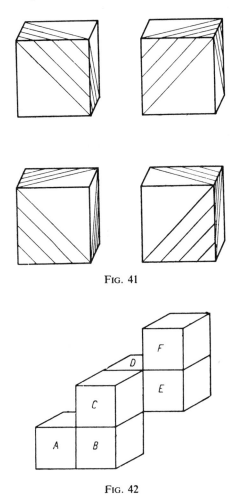

FIG. 41

FIG. 42

in the system of cubes A to F along a straight line. If the track of this motion is a closed hexagon, then the point on the front face A from which the molecule started must by repeated reflec-tions become identical with the point on the back wall F. If we

pass the line as shown in Fig. 43, then it will pierce all cubes
$A, ..., F$, without leaving these cubes. If we reduce six cubes
to one, by reflecting the cube F in the face which separates F from

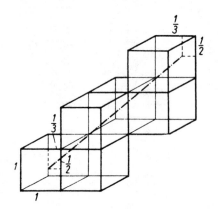

FIG. 43

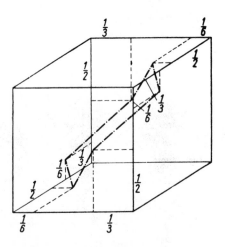

FIG. 44

E, and by reflecting the cube E in the face which separates it from
D, etc. then as the path of the movement we obtain a hexagon
without knots, as shown in the Fig. 44.

34. We give all possible diagrams (of which there are 11) as shown in Fig. 45. The first six solutions are those in which four faces of the cube are arranged into one strip. There are no other solutions of this kind. The next four diagrams are those in which there

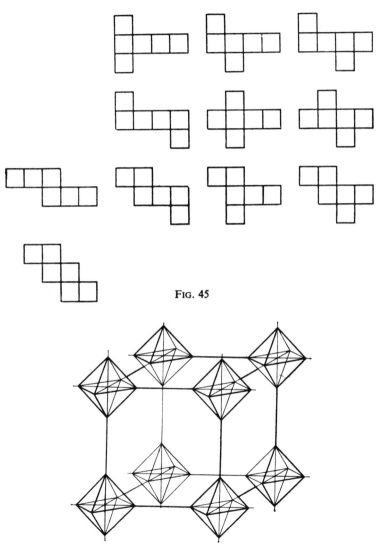

FIG. 45

FIG. 46

are three faces arranged into one strip, but there are no four such faces. And finally, in the last sketch, no strip has three faces.

35. The solids formed by cutting off eight corners have fourteen faces (the number of faces of the cube plus the number of corners of the cube). Eight of them are triangles, six are octagons (Fig. 46).

For the largest octahedra, the faces of the 14-hedron will become triangles and squares (Fig. 47). As the maximum octahedron arising by cutting off the eight corners of eight cubes is composed of eight pyramids whose heights are equal to half the length of the cube edge, the eight pyramids will constitute $8 \cdot \frac{1}{3} \cdot \frac{1}{8} \cdot \frac{1}{2} = \frac{1}{6}$ of the whole space. At every vertex there are joined six solids: four 14-hedra and two octahedra.

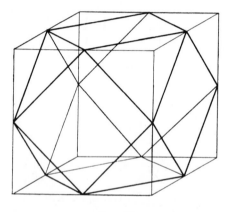

FIG. 47

36. It is not difficult to see that the answer to the question put in the problem is in the affirmative, the hexahedron which satisfies the conditions of the problem is a parallelepiped which has equal edges, and three equal plane angles at one of the corners.

Let us suppose that we have a given rhombus with an acute angle a and diagonals $2a$ and $2b$. Putting three such rhombi together by the vertices of acute angles, we obtain a three-faced corner. Both the corner itself and its rectangular projection on the plane (viewed from the vertex) is shown in Fig. 48. Two such corners put together give a hexahedron of the kind sought in the problem.

We see that if $\alpha < 60°$, i.e. if $b = a\cot\tfrac{1}{2}\alpha > a\sqrt{3}$, then the hexahedron can be constructed only in the described manner. If, however, $\alpha > 60°$, that is, if $a < b < a\sqrt{3}$, i.e. $a > b/\sqrt{3}$, then three rhombi can be combined to form a three-faced corner

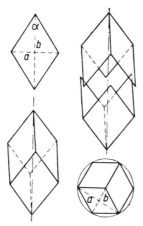

FIG. 48

not only by putting together the vertices of the acute angles but also by doing so with the vertices of the obtuse angles $180° - \alpha$ (Fig. 49). In this case, besides the hexahedron from Fig. 48, we obtain the hexahedron from Fig. 49 satisfying the condition of the problem.

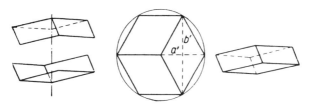

FIG. 49

If $\alpha = 90°$, then $a = b$ and both hexahedra are the same solid, namely a cube.

In Fig. 50 we have the diagrams of the two hexahedra considered in the problem.

It is worth-while to note that the hexahedra from Figs. 48 and 49 are examples of two different convex polyhedra with the same number of faces pair-wise equal.

37. The answer depends upon whether we consider two tetrahedra which are mirror images of one another as different or not.

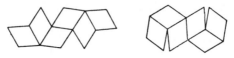

FIG. 50

We shall show that in the first case there are 60 different tetrahedra; in the second case there are obviously only 30.

We see in Fig. 51 a sketch of a tetrahedron, with edges denoted

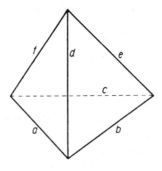

FIG. 51

by the letters a, b, c, d, e, f. The sticks from which we are to assemble tetrahedra are labelled with numbers from 1 to 6. A stick No. k can be placed instead of an arbitrary edge a, b, c, d, e, f. Thus there are $6! = 720$ possibilities, that is, as many as there are permutations of six objects.

However, not all such permutations will yield different tetrahedra. Some of them will be equal, but differently placed.

Let us suppose for a moment that the tetrahedron shown in Fig. 51 has all its edges of the same length. We shall investigate in

how many different ways this tetrahedron can be placed so that
its edges be covered with the edges of the initial position. We see
(1) that we can choose as a base any one of the four faces of the
tetrahedron, and (2) that the base is a triangle with all sides different, which can, of course, be placed in three different ways.
Thus we have altogether $3.4 = 12$ different positions for the same
tetrahedron, and they are all the possible different positions of
this tetrahedron, such that its edges coincide with the edges of the
original position. Thus the 720 possible permutations of the sticks
1–6, taken as edges a–f, generate every tetrahedron 12 times in different positions. We thus see that there are $720:12 = 60$ different
tetrahedra which can be assembled from 6 different sticks. If we
consider every two tetrahedra which are mirror-images of each
other as being identical, then there will be 30 different tetrahedra.

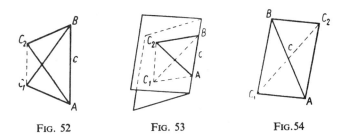

FIG. 52   FIG. 53   FIG.54

38. Let us take two triangles with the sides a, b, c and let us arrange them in such a way that the side c is common to both triangles
(Fig. 52). Unfolding the planes of these triangles (Fig. 53), we
increase the distance between the vertices C_1 and C_2 (Fig. 53)
up to the maximum distance when these triangles are again lying
in the same plane but their vertices are on opposite sides of the
line AB (Fig. 54).

Let us denote the distance $C_1 C_2$ in Fig. 52 by d_1 and the same
distance in Fig. 54 by d_2. The condition of the existence of a tetrahedron satisfying the conditions of the problem is the occurrence
of the inequality $d_1 < c < d_2$.

To satisfy the inequality $d_1 < c$ it is necessary and sufficient
that each of the angles A and B be acute, and in order that the

inequality $c < d_2$ be satisfied it is necessary and sufficient that angle C be acute. Thus the condition of the existence of the tetrahedron is that the triangle ABC have acute angles.

Let us suppose the above condition to be satisfied. In Fig. 55 we have a diagram of a tetrahedron satisfying this condition, and at the same time $A_1 B_1 \| AB$, $B_1 C_1 \| BC$, $C_1 A_1 \| CA$.

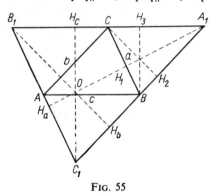

FIG. 55

Let $BC = a$, $CA = b$, $AB = c$, let $a \leqq b \leqq c$, and $2p = a+ +b+c$.

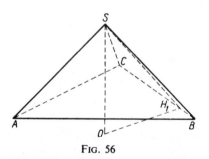

FIG. 56

Retaining the symbols used in Figs. 55 and 56, we see that the height $h = SO$ of the tetrahedron can be computed from the right triangle $SH_1 O$ if we first compute $SH_1 = A_1 H_1$ and $H_1 O$. In fact

$$A_1 H_1 = \frac{2}{a} \sqrt{p(p-a)(p-b)(p-c)},$$

$$OH_1 = H_1 H_a - OH_a, \qquad H_1 H_a = A_1 H_1;$$

on the other hand OH_a can be found from the proportion

$$\frac{OH_a}{C_1 H_a} = \frac{B_1 H_c}{C_1 H_c},$$

resulting from the similarity of triangles $C_1 OH_a$ and $C_1 B_1 H_c$. A simple calculation gives

$$OH_a = \frac{(a^2+b^2-c^2)(a^2+c^2-b^2)}{4a\sqrt{p(p-a)(p-b)(p-c)}}.$$

We now have

$$h^2 = (A_1 H_1)^2 - (H_1 H_a - OH_a)^2 = OH_a(A_1 H_a - OH_a)$$

$$= \frac{(a^2+b^2-c^2)(a^2+c^2-b^2)(b^2+c^2-a^2)}{8p(p-a)(p-b)(p-c)}.$$

Finally, denoting the volume of the tetrahedron by V, we have

$$V = \frac{h}{3}\sqrt{p(p-a)(p-b)(p-c)}$$

$$= \frac{1}{6\sqrt{2}}\sqrt{(a^2+b^2-c^2)(a^2+c^2-b^2)(b^2+c^2-a^2)}.$$

39. There exists a $2n$-hedron satisfying the conditions of the problem, and it is easy to give its construction. Let us imagine two parallel circular disks T_1 and T_2 (Fig. 57). Let us divide the perimeter of each disk into $n \geq 3$ equal parts; let P_1, P_2, ..., P_n and Q_1, Q_2, ..., Q_n denote the points dividing the perimeters of disks T_1 and T_2, respectively. Let us turn disk T_1 in its plane so that the perpendicular projection of point P_k upon the plane of disk T_2 falls on the centre of the arc $Q_k Q_{k+1}$ (the projection of P_n shall fall on the centre of the arc $Q_n Q_1$).

Let us lay off on the straight line $O_1 O_2$, on the opposite sides of the segment $O_1 O_2$, segments $O_1 S_1$ and $O_2 S_2$ of equal length (Fig. 57).

Then let us connect by segments the point S_1 with each of the points P_k and the point S_2 with each of the points Q_k. Similarly, let us join each point P_k with points Q_k, Q_{k+1}. In this way we obtain the skeleton of a $2n$-hedron, a fragment of which is shown in Fig. 57.

Let us show that we can choose the elements of this polyhedron in such a way as to obtain a $2n$-hedron with equal faces.

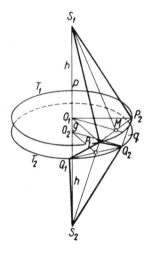

FIG. 57

In fact let the length of the segment $O_1 O_2$ be g. We consider the tetrahedron $S_1 P_1 Q_2 P_2$. Let M denote the centre of the segment $P_1 P_2$. In order that the polyhedron mentioned above be a $2n$-hedron with congruent faces, it is necessary and sufficient that the segment $S_1 O_2$ intersect the segment $P_1 P_2$ at the point M. The condition will be satisfied when the following equality occurs: $h : O_1 M = (h+g) : O_2 Q_2$; hence

$$g = O_1 O_2 = h \, \frac{O_2 Q_2 - O_1 M}{O_1 M} \, .$$

For $n = 4$ we obtain the required octahedron.

We can show that the faces of the $2n$-hedron satisfying the conditions of the problem are congruent deltoids, which are rhombuses only for $n = 3$. (See problem 36.)

40. Let the base of the pyramid be an equilateral triangle ABC, and let the lateral faces be isosceles triangles with top angles equal to 30°. In Fig. 58 we see the diagram formed by intersecting the pyramid along the edge CT, and then unfolding it on the plane.

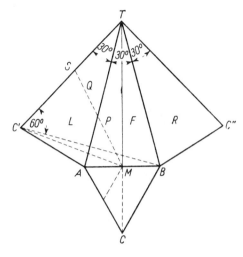

FIG. 58

The distance from the midpoint M of AB to the top T of the pyramid is given by the length of the straight segment MT on the diagram. This follows immediately from the fact that MT is straight, not only on the diagram but also in space, whereas all the other possible paths leading from M to T on the surface of the pyramid are not straight, and we know that any path connecting two points and different from a straight segment connecting them is longer than that segment. However, we still have to show that the point T is furthest from the point M. The triangle $TC'B$ is equilateral, and thus the angle $TC'B$ is 60° and the angle $TC'M$ is more than 60°. As the angle $C'TM$ is 45°, this angle is less than the angle $TC'M$, and thus the side TM is longer than the side $C'M$. It follows that if S runs along the side $C'T$ then the length of the segment MS increases from the length MC' to the length MT. It appears that to every point S of the edge $C'T$ there leads from M a shorter path than to the top. But, as shown in Fig. 58, the point Q on the face L can be reached earlier than S on the path $MPQS$. From this it follows that to every point Q of this face there leads from M a way shorter than the segment MT which is the shortest path from M to T. Thus all the points Q of the face L lie nearer to the point M than T, and the same holds for the face R.

We immediately see that on the face F the furthest point from M is the point T. And on the base ABC the point C, furthest from M, is evidently nearer to M than the point T. Thus we have shown that the point T is furthest from M, and all the other points in the tetrahedron are nearer.

We advise the reader who wishes to understand this answer thoroughly to reflect why we did not use a regular tetrahedron as an example.

41. The polyhedron can be considered as a spatial network whose sides are the edges, nodes are the vertices, and meshes are the faces of the polyhedron. The path of the fly ought to form a closed polygon, without multiple points, and belonging to the above-

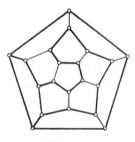

FIG. 59

mentioned network. The possibility of defining such a path still remains if we spread out the network in order to put it on the plane.

In Fig. 59 there is a network of a regular dodecahedron spread-out on the plane. The heavy line represents the path of the fly satisfying the conditions of the problem.

Let us spread out on the plane, in a similar way, a network of a rhombic dodecahedron (Fig. 60). The nodes of the network can be divided into 2 classes: those which are joints of 3 sides, and those which are joints of 4 sides (black points in the figure).

Every node of the first class is connected by segments with nodes of the second class only, and *vice versa*. Thus the fly in its roaming would be compelled to pass through the nodes of the

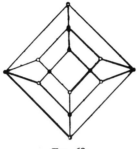

FIG. 60

first class and the nodes of the second class alternatively. As there are 8 nodes of the first and 6 nodes of the second class, the fly cannot visit all the vertices of the rhombic dodecahedron by walking along the edges so as not to pass by any vertex twice and return to the starting point.

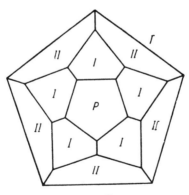

FIG. 61

42. Let us place a regular dodecahedron in such a position as to have one of its faces P before us (Fig. 61). We shall call this face the front face. The invisible face parallel to it may be called the back face. The set of five visible faces surrounding the first face is called ring *I*, and the set of five invisible faces surrounding the back face is called ring *II*.

As the essence of this problem is to establish the number of possible ways of painting the faces of a dodecahedron, we can replace the *n*-hedron by the plane representation in Fig. 61 (to its back face *T* corresponds the part of the plane lying outside the big pentagon) or by an even simpler representation given in Fig. 62.

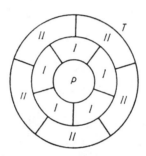

FIG. 62

It is impossible to paint the regular dodecahedron with three colours in such a way as to have neighbouring faces of different colours. In fact, if, for instance, the front face is of colour *A*, then the five faces of ring *I* connected with the front face ought to be painted with colours *B* and *C*, which is impossible.

Let us suppose that the faces of the dodecahedron can be painted in the way required in the problem with four colours: *A*, *B*, *C*, *D*. It is easy to see that each of these colours would appear three times. To prove it, let us assume the contrary. If a colour, for instance colour *A*, appears less than three times, then another colour, for instance *B*, would occur more than three times. Let us suppose that, for instance, the front face is of colour *B*. Then in ring *I* colour *B* could not appear; consequently, of the six invisible faces at least three should be of colour *B*; then in ring *II* colour *B* could not appear at all, and if the back face were not of colour *B*, then in ring *II* colour *B* could appear at most twice.

From the above reasoning it follows that the back face cannot be of the same colour as the front one, but that it must be of a colour which appears twice in ring *I*.

We see, finally, that by choosing the colours of ring *I*, the order in which these colours appear, and the colour of the opposite face, we already determine uniquely the way of painting the whole solid. Indeed, in the case shown in Fig. 63, for instance, the front face

FIG. 63

has to be of colour *A*; as regards the faces of ring *II*, face *1* cannot differ from *A*; consequently face *2* is of colour *D* and face *5* of colour *B*, whence faces *3* and *4* are of colours *B* and *A*, respectively. This example proves that a dodecahedron can be painted with four colours.

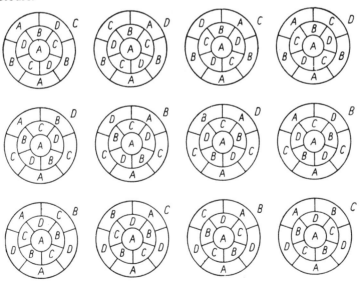

FIG. 64

Let us suppose that the front face of a dodecahedron painted with four colours is of colour A. Then in ring I there can appear six different combinations of colours, and since in every combination of colours the back face can be painted with two different ways, we have 12 possible combinations of colours, as shown in Fig. 64. However, on a dodecahedron painted with four colours, colour A appears on three faces only; thus, if we choose any of the ways of painting shown in Fig. 64, then, by rotating the dodecahedron, we can exhaust at most two more ways of painting, which makes altogether three ways. From this it follows that there are at least four ways of painting a dodecahedron.

It is easy to prove that there are exactly four ways of painting.

In fact, by each of the patterns shown in the first row of Fig. 64 we have exhausted those ways which are shown in the same column

43. We take an arbitrary convex polyhedron. We assume one of its faces to be the base of a pyramid whose bihedral angles at the base are so small that the polyhedron obtained from the convex polyhedron by superposition of the pyramid upon it is still convex, and that it is possible to hollow out in the original polyhedron

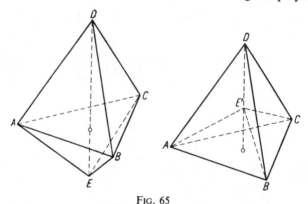

Fig. 65

a pyramid symmetric to the given pyramid with respect to a chosen face. In this way we obtain two polyhedra, one convex and the other non-convex with convex and pair-wise equal faces. In Fig. 65 we see two polyhedra $ABCDE$ and $ABCDE'$, obtained in the above manner from the tetrahedron $ABCD$.

Two 30-hedra, one convex and the other non-convex with convex and pair-wise equal faces, can be seen in Figs. 66 and 67.

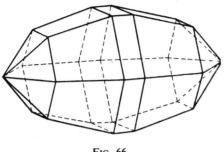

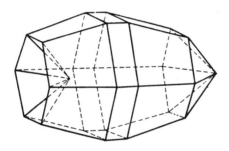

FIG. 66

FIG. 67

44. Examples of solids satisfying the conditions of the problem are shown in Figs. 68 and 69. The solid in Fig. 69 consists of two

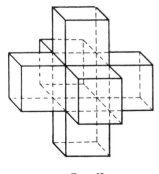

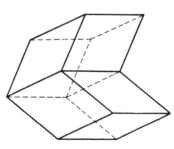

FIG. 68

FIG. 69

parallelepipeds with rhombic faces. These parallelepipeds are adherent, having one common face.

There also exists a non-convex rhombic dodecahedron. It can be obtained from a convex rhombic dodecahedron if we remove the front three-faced corner (see Fig. 70, left side), and replace

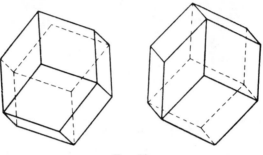

FIG. 70

it by the back three-faced corner shifting it by parallel motion along the edges formed by the remaining six faces (see Fig. 70, right side).

We see that the newly formed solid consists of three parallelepipeds. If we remove one of them, we shall obtain the solid, shown in Fig. 69.

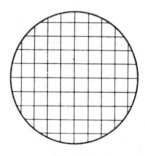

FIG. 71

45. Three models of conical planets discussed in this problem are obtained from a rectangular network of "meridians" and "parallels" (Fig. 71) cutting it in one of the three ways shown in

Fig. 72, and winding up the remaining part into a cone with top N.

Figure 73 shows us the top view. We see distinctly two families of curves — "meridians" and "parallels". Neither the "meridians" nor the "parallels" intersect each other, but every meridian cuts

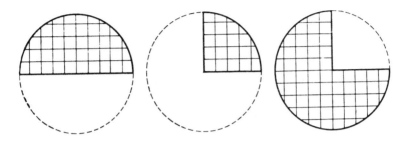

FIG. 72

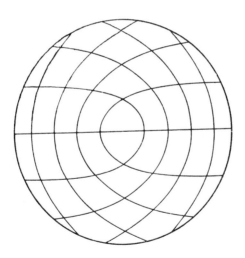

FIG. 73

every parallel at two points, as on the globe. The shortest paths have a constant direction, that is, they are intersecting meridians and parallels with a constant angle.

In Figs. 74 and 75 we see the second and the third cone from the top view, that is, the projection of families of curves on a plane

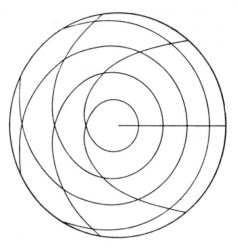

FIG. 74

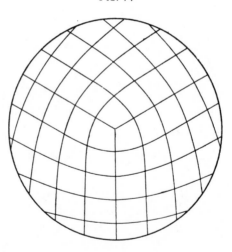

FIG. 75

perpendicular to the axes of the cones. On the first cone we have one family of curves, and on the second three families. It is of interest to note that the solutions given here are not unique.

46. Of three given spheres no two can be mutually tangent because the plane passing through the tangent point P would be tangent to both spheres, and, moreover, would have a common circle with the third sphere. The line tangent to this circle at the point P would thus be tangent to three spheres, in contradiction to the assumption.

Thus two spheres must have a common circle, on which the point P obviously lies. This circle passes through the third sphere, for otherwise the line tangent to the circle at the point P would again be the common tangent of the three spheres. As this circle passes through the third sphere at the point P, it must pass through it at an additional point which lies on all three spheres.

47. Let the circle K be a section of the surface S referred to in the problem. Let L be the axis of the circle K, i.e. the straight line leading through the centre of the circle K perpendicular to the plane which gave K as a section. Let P be a certain plane passed through L. From the above assumption it follows that P cuts out a circle from S, and as P intersects K at two points, A and B, so that AB is the diameter of the circle K, the circle cut out by P, let us call it $C(P)$, has a chord AB, the axis L being the line of symmetry of this chord. As we know, the line of symmetry of a chord of a circle passes through its centre, and thus L passes through the centre of the circle $C(P)$ and cuts this circle at the points M and N lying on the axis L and also on the surface S. Thus every circle $C(P)$ which arose from passing an arbitrary plane P through L must contain points M and N because they lie on both S and P. There are no other points on L belonging to S, because if Q were such a point then the section P would contain 3 different points M, N, Q lying on the straight line L and on S, and thus the straight line L would cut the circle $C(P)$ at 3 points, which is impossible. We see from this that the segment MN on L is the common diameter of all circles $C(P)$. From this it follows that these circles can be obtained from one of them by rotating it about the common diameter MN. This gives, of course, a sphere whose axis is MN. So the set of all sections made by L gives us the sphere Σ as a part of the surface S. But S does not contain anything else besides Σ, because if a certain point T not belonging to Σ were lying on S, then a straight line through T and through the centre of the sphere Σ would cut two points, T_1 and T_2, on Σ and we would have on S the three points T, T_1, T_2 on one straight line, which is impossible. Thus S is identical with the sphere Σ.

48. Electric light in the majority of cities is supplied by alternating current of 50 cycles; in one second an electric lamp lights up and goes out 100 times, and between two successive moments of maximum current there always elapses 0·01 second.

If the disk rotates clockwise with a speed of 25 revolutions per second, then during a period of 0·01 second each of the sections of the interior disk will perform 0·25 revolutions, and will take the same position as that taken previously by the section of the same colour. Because of this, it would seem to us that the inside disk does not rotate. With the rotating speed slightly exceeding 25 revolutions per second, we shall have the impression that the disk is rotating in a clockwise direction, and with a rotating speed slightly less than 25 rev/sec, it will seem to us that the disk rotates counterclockwise.

For the exterior ring things are similar, but the critical speed is 20 rev/sec.

Let us now suppose that we make the disk rotate very fast. When the decreasing speed reaches 25 rev/sec it will seem to us that the inside disk is motionless and the outside ring rotates clockwise. With the speed of the disk less than 25 rev/sec but more than 20 rev/sec we shall have the impression that the inside disk rotates in the direction opposite to that of the outside ring. With 20 rev/sec the external ring will stop, and later the inside disk as well as the outside ring will both rotate counterclockwise.

The toy functions best with fluorescent light, and badly with ordinary electric light bulbs of high voltage, whose filament has a high thermal inertia and does not give a strong contrast during one cycle.

49. The dispute can be settled in the following manner: We give the priority of choosing the piece of ham to the third co-owner. She will choose, of course, the piece which according to her home

balance is not less than either of the remaining two pieces. That is
the piece whose value, according to her opinion, is not less than
$ 4·00. Such a piece must exist because, by division of the whole
into 3 parts, one of the parts cannot be less than 1/3 of the total
weight.

Afterwards the second woman chooses her piece. She must
also be satisfied because, after the third woman took her share,
there remained at least one piece which, according to the balance
in the shop on the corner, corresponded to a value not less than 1/3.

The first woman, who receives the remaining piece, must be
satisfied, since she considered all the pieces to be of equal weight.

50. Let us call a system of two perpendicular straight lines a cross.
In Fig. 76 we see such a cross marked with dashed lines — its

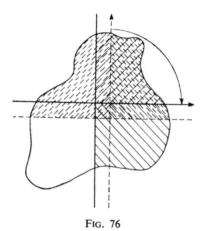

FIG. 76

vertical arm bears an arrow directed upwards. Whatever the shape
of the region, the cross can always be moved, without turning,
in such a way that the two parts of the plane region adjacent to
the line distinguished by the arrow and lying above the horizontal
dashed line, will have areas equal to $P/4$, P being the area of the
whole region; these upper quarters are shaded with dashed lines.
If the lower quarters also have equal areas, then the theorem is
proved.

Let us assume the contrary: e.g. that the left lower quadrant has an area greater than $P/4$ (thus greater than that of the right lower quadrant). We turn the cross to the right in such a way that the quadrants adjacent to the selected arm continue to have areas equal to $P/4$. After a rotation of 90° the cross will assume the position denoted in Fig. 76 by the continuous line. The vertical arm has to move to the left, and the horizontal arm upward, in order that the shaded quadrants (with continuous lines) have areas equal to $P/4$. In this new position of the cross the left lower quadrant (with respect to the selected arm) has an area less than $P/4$ since it is a part of the dashed quadrant whose area is $P/4$, and thus the right lower quadrant has an area $P/4$. This new position of the cross gives a different quartering from the former one, which implies that during the rotation there was a moment at which all quadrants had areas equal to $P/4$.

The proof of the feasibility of quartering the pie presented here has not required calculations. But it would require calculations (and not quite easy calculations!) to obtain an effective quartering of a given triangle, for instance with sides of length 3, 4, 5.

51. We see that if at first Ted chooses as the point P the centre of gravity of the triangle (Fig. 77), then the best division of the

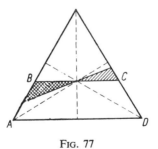

FIG. 77

pie for Ned is when the cut is parallel to one of the sides of the triangle. The proof is to show that the triangle shaded with dashes in Fig. 77 has an area less than the triangle twice shaded. The area of the part $ABCD$ of the pie is equal to $5/9$ of the area of the whole pie; thus the surplus is given by the ratio $5:4$.

It is easy to see that the choice of the centre of gravity is the best one for Ted, because in every other position of the point *P* Ned would be able to cut for himself not only a trapezoid *ABCD*, but an additional piece of pie as well (Fig. 78).

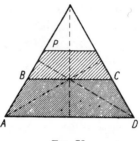

Fig. 78

We leave it to the reader to answer the remaining questions· At any rate it is easy to show that if the pie has a very strange shape, such as is shown in Fig. 79, then for every choice of the

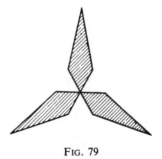

Fig. 79

point *P* Ned can cut at least 2/3 of the pie. Finally we propose to the reader to prove by the use of the result of the former problem (quartering of the pie) that no shape of the pie makes it possible for Ned to cut more than 3/4 of the area.

52. We compare the first pair, then the second one, and afterwards we compare the object which appeared to be heavier in the first pair, with the heavier object from the second one. We can write the result of these three weighings as

$$A < B < C$$
$$\vee$$
$$D$$

where $M < N$ is to be read: "M is lighter than N". The fifth object E can be ranged in the row ABC by comparing it first with B; if it appears to be heavier than B, we shall compare it with C, if it appears to be lighter than B, we shall compare it with A. In such a way two weighings will place E among ABC, and will lead to one of the following configurations:

(1) $\quad A < B < C < E,$ $\qquad$ (2) $\quad A < B < E < C,$
$\qquad\qquad\vee$ $\qquad\qquad\qquad\qquad\qquad\vee$
$\qquad\qquad D$ $\qquad\qquad\qquad\qquad\qquad D$

(3) $\quad E < A < B < C,$ $\qquad$ (4) $\quad A < E < B < C.$
$\qquad\qquad\vee$ $\qquad\qquad\qquad\qquad\qquad\vee$
$\qquad\qquad D$ $\qquad\qquad\qquad\qquad\qquad D$

Up to now we have made altogether five weighings. If they have resulted in (1) we compare D with A; if it appears that $D < A$ then this sixth weighing gives already the result sought for; if, however, it appears that $A < D$ then we compare D with B, and this seventh weighing will terminate the procedure, yielding $A < D < B < C < E$ or $A < B < D < C < E$. If the first 5 weighings resulted in (2), then we should range D in the row ABE by means of two weighings, starting with the comparison of D with B — altogether seven weighings again. Configurations (3) and (4) differ from configuration (2) only in the labelling of the objects; we shall treat them in the same way as the configuration (2).

Thus seven weighings always suffice to range 5 objects.

53. Let us denote successively the days of the week by I, II, ..., VII, and let us make a table assuming that the day of the week which corresponds to a certain date of 1911 was denoted by number I.

A	B	C	D	A	B	C	D
1911	I	I		1931	V	V	
1912	II	III	I	**1932**	VI	VII	V
1913	IV	IV		1933	I	I	
1914	V	V		1934	II	II	
1915	VI	VI		1935	III	III	
1916	VII	I	VI	**1936**	IV	V	III
1917	II	II		1937	VI	VI	
1918	III	III		1938	VII	VII	
1919	IV	IV		1939	I	I	
1920	V	VI	IV	**1940**	II	III	I
1921	VII	VII		1941	IV	IV	
1922	I	I		1942	V	V	
1923	II	II		1943	VI	VI	
1924	III	IV	II	**1944**	VII	I	VI
1925	V	V		1945	II	II	
1926	VI	VI		1946	III	III	
1927	VII	VII		1947	IV	IV	
1928	I	II	VII	**1948**	V	VI	IV
1929	III	III		1949	VII	VII	
1930	IV	IV		1950	I	I	

In column *A* years are given (leap years are printed in bold type). Column *B* contains days of the week corresponding in different years to the same date as in the year 1911 if this date is in the interval 1 January – 28 February. Column *C* contains days of the week corresponding in different years to the same date if this date is in the interval 1 March – 31 December. Column *D* contains days of the week corresponding in particular leap years to the date of 29 February. We see from this table that the day of the week corresponding to any fixed date before 1 March shifts after an ordinary year by 1, and after a leap year by 2; but the day of the week corresponding to any date after 1 March shifts by 1 before an ordinary year, and by 2 before a leap year.

We also see from this table that if a given date (different from 29 February) corresponded in a certain year, say, to Monday, it will fall again on the same day of the week either after 5 years, or after 6 years, or after 11 years (under the condition that in this period there has been no year whose number is divisible by 100 but not divisible by 400).

If Mrs. Z. had been born on any other day than 29 February, and up to 27 July 1950 celebrated her birthday only once, then on that day she would still not be marriageable and there would be no reason to address her as Mrs. Z. or to describe her as "not yet old".

Let us suppose then that Mrs. Z. was born on 29 February. The day of the week corresponding to this date shifts every 4 years by 5 days, and thus 29 February falls on the same day of the week every $7.4 = 28$ years (if in this period there is no year divisible by 100 and not divisible by 400). If Mrs. Z. "is only one year old" then she celebrated her birthday not earlier than in 1924 and not later than in 1948.

Since we know that Mrs. Z. was born after World War I, she celebrated her first birthday in the year 1948, and she was born on 29 February 1920.

54. Let n denote the number of fish in the pond that are suitable to be caught. Then the ratio of the number of marked fish to the total number of fish is $30/n$.

The second time the ichthyologist caught 40 fish, of which two were marked. The ratio of the number of marked fish to the total number of fish is $1/20$.

If we assume that the marked fish were equally distributed in the pond among other fishes, the above ratios must be equal; thus $30/n = 1/20$, whence $n = 600$.

55. In every case four tests will suffice if we assume the following method: the roller is first compared with the middle hole, that is, the eighth in the row. Then, depending on the result, either with the fourth or with the twelfth, etc. The result of each test is the answer "yes" (if the roller can be inserted into the hole), or "no" (if the roller cannot be inserted there). Thus four tests give 16 possibilities, that is, as many as there are different diameters distinguishable with the instruments.

56. If $2n$ numbers $x_1, x_2, \ldots, x_{2n}$ are arranged in increasing order and if $x_m < x < x_{m+1}$, then the sum of absolute errors when we replace the numbers $x_1, x_2, \ldots, x_{2n}$ by x is equal to

$$(x-x_1)+(x-x_2)+\ldots+(x-x_m)+(x_{m+1}-x)+\ldots+(x_{2n}-x). \quad (1)$$

It is easy to prove that the sum (1) is the least when we write for x a number y, where $x_n < y < x_{n+1}$:

$$(y-x_1)+(y-x_2)+\ldots+(y-x_n)+(x_{n+1}-y)+\ldots+(x_{2n}-y). \quad (2)$$

In fact, if for instance $m > n$ (i.e. $x > y$) then the difference between (1) and (2) is equal to

$$n(x-y)+\big((x-x_{n+1})+\ldots+(x-x_m)\big)-$$
$$-\big((x_{n+1}-y)+\ldots+(x_m-y)\big)+(2n-m)(y-x)$$
$$> n(x-y)+0-(m-n)(x-y)+(2n-m)(y-x)=0.$$

Similarly, it is possible to prove that if we divide into two parts a set of $4n$ numbers $x_1, x_2, \ldots, x_{4n}$,

$$x_1, x_2, \ldots, x_{2m} \quad \text{and} \quad x_{2m+1}, x_{2m+2}, \ldots, x_{4n},$$

and then replace the numbers of the first part by a number x' such that $x_m < x' < x_{m+1}$, and the numbers of the second part by a number x'' such that $x_{2n+m} < x'' < x_{2n+m+1}$, the sum of absolute errors will be the smallest when $m = n$.

Hence is follows that the limiting diameter of 120 ball-bearings is 6·09 mm (there are 60 ball-bearings of diameter smaller than 6·09 mm and as many ball-bearings of diameter greater than 6·09 mm) and the inscriptions on the boxes are: $a = 6·06$ mm; $b = 6·15$ mm.

57. The area of the section of the roll is 25π cm², of which the ribbon takes up 25 cm²; thus the interior has an area of $25(\pi - 1)$ cm². Let us denote by d the diameter of the roll (without ribbon); from the equation

$$\pi \frac{d^2}{4} = 25(\pi-1)\,\text{cm}^2$$

it follows that

$$d = 10\sqrt{\frac{\pi-1}{\pi}}\,\text{cm} \approx 8·26\,\text{cm}.$$

58. As is well known, the time shown by the clock is completely defined by the coordinate of the hour hand on the scale of the clock (Fig. 80); the minute hand has only a secondary rôle, namely it

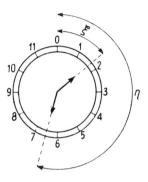

FIG. 80

forms together with the corresponding scale on the face of the clock a kind of vernier which enlarges twelve-fold the accuracy of the reading. If we denote the coordinate of the hour hand on the face of the clock by ξ, and the coordinate of the minute hand by η, we shall have

$$\eta - 12\{\xi\} = 0, \qquad 0 \le \xi < 12, \qquad 0 \le \eta < 12,$$

where $\{\xi\}$ denotes the fractional part of the number ξ.

Let us suppose that the two hands of the clock are equal. Let us denote by x the coordinate of one hand, and by y the coordinate of the second one. Three cases can occur:

    I. If $x - 12\{y\} \ne 0$, then $y - 12\{x\} = 0$.

    II. If $y - 12\{x\} \ne 0$, then $x - 12\{y\} = 0$.

    III. If simultaneously the equations

$$y - 12\{x\} = 0 \quad \text{and} \quad x - 12\{y\} = 0$$

occur, then the clock with equal hands cannot tell the time: as each hand can be taken as the hour hand, time x is just as likely as time y. There are 143 such exceptional positions of the hands of the clock.

The numbers x and y corresponding to them are coordinates of 143 points of intersection of the graphs of the functions

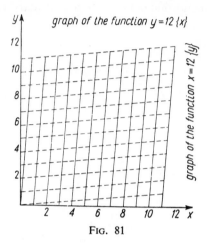

FIG. 81

$y = 12\{x\}$ and $x = 12\{y\}$ (Fig. 81). These points lie on the 23 straight lines defined by the equations

$$y = x + \frac{12}{13}k \quad (k = 0, \pm 1, \ldots, \pm 11). \tag{1}$$

If we now denote by

$$r = |x - y|$$

the error which we make by taking instead of the indication x of the clock the indication y, or *vice versa*, then according to equation (1) we shall obtain

$$r = \frac{12}{13}|k| \quad (k = 0, \pm 1, \ldots, \pm 11).$$

Rejecting, according to the conditions of the problem, errors greater than six hours, we obtain for $k = \pm 6$ the greatest value for the error. It is

$$r = \frac{72}{13} = 5\frac{7}{13},$$

i.e. 5 hours, 32 minutes and $18\frac{6}{13}$ seconds. (At what time of the day does this error threaten us?)

Obviously in our reasoning we have assumed that the owner of the clock reads the coordinates of the hands without error.

59. 1. It could happen that the same boy was the shortest giant and the tallest midget. The class consisting of km pupils of different height (k and m are natural numbers greater than 1, k denotes the number of columns and m denotes the number of rows) can be arranged in a rectangular manner in such a way that an arbitrary pupil p who has at least $k-1$ colleagues shorter than himself, and at least $m-1$ colleagues taller than himself, is at the same time the shortest giant and the tallest midget:

2. There are no classes in which the shortest giant is shorter than the tallest midget.

To prove it, let us denote by p the tallest midget and by P the shortest giant. Let us suppose that $P < p$. Pupils p and P can stand neither in the same column, because then p would not be the shortest in his column, nor in the same row, because then P would not be the tallest in his row. Let us denote by p_{ik} the pupil standing on the intersection of that column and that row in which pupils p and P are standing. We have

$$p < p_{ik}, \quad p_{ik} < P;$$

thus $p < P$, in contradiction to the assumption made above that $P < p$. The supposition $P < p$ leads to a contradiction, and thus it is false.

3. If the teacher, when choosing the giants, looked for them in the same way as for the midgets, — i.e. in columns and not in rows — then, of course, no pupil could be both the shortest giant and the tallest midget, because he would have to be, at the same time, the shortest pupil and the tallest one in his column, which is not possible.

In this case, however, the shortest giant can be either taller or shorter than the tallest midget.

In fact, let us choose quite arbitrarily from a class consisting of km pupils of different height $2k+m-2$ pupils and let us place them along 3 sides of a rectangle in decreasing order with respect to height,

$$p_1, p_2, \ldots, p_{2k+m-2}$$

in the following way:

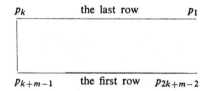

We arrange the remaining pupils quite arbitrarily in a rectangular manner.

Pupil p_k is the shortest giant, pupil p_{k+m-1} is the tallest midget, and as $p_k > p_{k+m-1}$, the shortest giant is taller than the tallest midget. If, on the other hand, $k+2m-2$ pupils $p_1, p_2, \ldots, p_{k+2m-2}$ are arranged in decreasing order with respect to height, the remaining pupils being arranged in a rectangular manner quite arbitra-

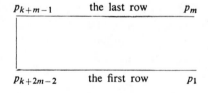

rily, then pupil p_{k+m-1} will be the shortest giant, and pupil p_m will be the tallest midget, and in view of the inequality $p_{k+m-1} < p_m$ the shortest giant will be shorter than the tallest midget.

60. Let us denote the items of the problem by the numbers they are designated with, and let the symbol $p \to q$ denote that the answer "yes" to question p implies the answer "yes" to question q. Then the following assertions are true:

$$1 \to 2, \quad 1 \to 3, \quad 1 \to 4, \quad 1 \to 7, \quad 1 \to 8, \quad 1 \to 10, \quad 1 \to 11, \quad 1 \to 12,$$
$$1 \to 13, \quad 2 \to 4, \quad 3 \to 7, \quad 4 \to 2, \quad 5 \to 7, \quad 6 \to 2, \quad 6 \to 4, \quad 6 \to 9,$$
$$7 \to 3, \quad 8 \to 3, \quad 8 \to 7.$$

Thus the answers to questions 2 and 4 are identical and the same is true for the answers to 3 and 7.

61. The symbols we speak about in the problem are the following:

(1)	$a\,b\,c\,d$	(9)	$a\,b'c\,d'$
(2)	$a\,b\,c\,d'$	(10)	$a'b\,c'd$
(3)	$a\,b\,c'd$	(11)	$a'b\,c\,d'$
(4)	$a\,b'c\,d$	(12)	$a\,b'c'd'$
(5)	$a'b\,c\,d$	(13)	$a'b\,c'd'$
(6)	$a\,b\,c'd'$	(14)	$a'b'c\,d'$
(7)	$a\,b'c'd$	(15)	$a'b'c'd$
(8)	$a'b'c\,d$	(16)	$a'b'c'd'$

We shall show that among these 16 symbols there are 2, namely $a\,b'c'd$ and $a'b\,c\,d'$, for which there are no adequate trains.

Let the symbol P_p denote the number of smoking passengers in the compartments for smokers, let P_n denote the number of smoking passengers in the compartments for non-smokers, let N_p denote the number of non-smoking passengers in the compartments for smokers, and N_n the number of non-smoking passengers in the compartments for non-smokers.

Let us now consider the assertion $a\,b'c'd$. Because of the meaning of the letters a, b', c', d the following inequalities

$$P_p > P_n, \quad N_p < N_n, \quad P_p < N_p, \quad P_n > N_n \qquad (1)$$

are true if $a\,b'c'd$ is valid.

From the first three inequalities of this system we obtain

$$N_n > N_p > P_p > P_n.$$

Thus $N_n > P_n$, which contradicts the fourth of the inequalities (1).

Thus the symbol $a\,b'c'd$ is contradictory and there is no train to which it applies.

The symbol $a'b\,c\,d'$ is also contradictory, for among the inequalities it denotes,

$$P_p < P_n, \quad N_p > N_n, \quad P_p > N_p, \quad P_n < N_n,$$

the first three contradict the fourth.

With the remaining 14 symbols we can denote real trains, as the following table indicates.

No. of the train	Symbol	Distribution of the passengers in the compartments			
		smokers		non-smokers	
1	$a\ b\ c\ d$	PPP	NN	PP	N
2	$a\ b\ c\ d'$	PPPP	NNN	P	NN
3	$a\ b\ c'\ d$	PPP	NNNN	PP	N
4	$a\ b'\ c\ d$	PPPP	N	PPP	NN
5	$a'\ b\ c\ d$	PPP	NN	PPPP	N
6	$a\ b\ c'\ d'$	PP.	NNN	P	NN
8	$a'\ b'\ c\ d$	PP	N	PPP	NN
9	$a\ b'\ c\ d'$	PP	N	P	NN
10	$a'\ b\ c'\ d$	P	NN	PP	N
12	$a\ b'\ c'\ d'$	PP	NNN	P	NNNN
13	$a'\ b\ c'\ d'$	P	NNNN	PP	NNN
14	$a'\ b'\ c\ d'$	PP	N	PPP	NNNN
15	$a'\ b'\ c'\ d$	P	NN	PPPP	NNN
16	$a'\ b'\ c'\ d'$	P	NN	PP	NNN

62. Let us arrange the symbols denoting the blood groups in rows and columns: we place donors in the columns and recipients in the rows,

		Recipients			
		O	A	B	AB
	O	+	+	+	+
Donors	A	−	+	−	+
	B	−	−	+	+
	AB	−	−	−	+

and in the table made in this way we put + or — signs on the intersections of rows and columns. We read this table choosing an arbitrary row (for instance the second one) and an arbitrary column (for instance the third one); if there is a + sign at the intersection then the donor of the chosen row can give his blood without any danger to the recipient in the chosen column; he cannot do so if there is a — sign at the intersection; the example of the second row and the third column shows that A cannot give his blood to B. Thus the table defines who can and who cannot give his blood to whom. We can verify that this definition satisfies law I. In fact this law requires that on the diagonal there should appear only + signs, and requires nothing more; thus it is satisfied. Law II requires that in the row of donor O there should appear only + signs; the table satisfies this law. Law III requires that in the column of recipient AB there should appear only + signs: this law, too, is satisfied by the table. If, in the relations I, II, III, we substitute successively for X the symbols O, A, B, AB, we shall obtain 9 and only 9 relations, namely those denoted on the table by + signs. All other positions (there are 7 of them) are filled with — signs. Minuses are in those and only those positions to which the laws I, II, III do not give the sign + by any substitution, and this is exactly the fulfilment of law IV. It appears that the table is equivalent to laws I–IV and this gives assertion (1). Assertion (2) can also be verified by the table. For instance: $O \rightarrow A$, $A \rightarrow AB$; therefore $O \rightarrow AB$, and the table confirms it; the assertion can be verified in the same manner in all cases. Assertion (3): From the fact that we cannot get $A \rightarrow B$ by any substitution of group symbols for X it follows according to IV that $A \rightarrow B$ is false.

63. Let us denote the brothers by the group symbols X and Y. As X cannot give his blood to his brother, in the table given in the solution of the former problem, at the intersection of the X-row with Y-column there is a minus sign. Since Y cannot give his blood to X at the intersection of the Y-row with the X-column there is also a minus sign. These two minus signs are symmetric with respect to the diagonal of the table, consisting of plus signs only, and such a pair of symmetrically placed pluses is to be found only at the intersections A-row $\times$ B-column, B-row $\times$ A-column. Therefore one of the brothers is A and the other is B. The table shows that A can take blood from O and A, B can receive blood from O and B. Since both brothers can receive blood from their mother, she must be O. The laws of inheritance require the ascribing of O to the one-letter symbols of the parents; let us then denote the mother by the symbol OO. The law of taking the letters from both parents shows that the brothers got their symbols A and B from their father, as the mother does not possess them. Thus the father is AB. Parents with symbols AB, OO can only have children with symbols AO and BO, that is A and B (according to the law of omitting O wherever it occurs); consequently the sister of the two brothers is either A or B. Thus according to the table she can give her blood to one and only one of her brothers. (Nobody in this family can take blood from the father, but everyone can take blood from the mother.)

64. If the pole M is situated between poles L and N, then the path

$$ABC...LMN...T$$

is shorter than the path

$$ABC...LN...TM$$

by the segment TM. Hence the conclusion: in order that the path

$$ABC...LMN...T$$

be the longest, it is necessary that the walker should always change his direction after having driven a nail. Then the path is expressed by the formula

$$d = \overline{AB} - \overline{BC} + \overline{CD} - \overline{DE} + ...;$$

if the number of poles is odd, we have

$$d' = \overline{AB} - \overline{BC} + \overline{CD} - \overline{DE} + ... + \overline{RS} - \overline{ST};$$

if the number of poles is even, we have

$$d'' = \overline{AB} - \overline{BC} + \overline{CD} - \overline{DE} + \ldots - \overline{RS} + \overline{ST}.$$

Let us denote by

$$x_0, x_1, x_2, \ldots, x_n \tag{1}$$

the coordinates of poles $A, B, C, \ldots, T$.

Considering what we have mentioned above, we get

$$x_0 < x_1, \quad x_1 > x_2, \quad x_2 < x_3, \quad x_3 > x_4, \quad \ldots$$

The assumption that the intervals between the poles are equal implies that sequence (1) can be considered as a certain permutation of numbers $0, 1, 2, \ldots, n$.

Then from the equalities

$$\overline{AB} = x_1 - x_0, \quad \overline{BC} = x_2 - x_1, \quad \overline{CD} = x_3 - x_2, \quad \ldots,$$

$$x_0 + x_1 + x_2 + \ldots + x_n = \frac{n(n+1)}{2},$$

we obtain

$$d' = n(n+1) - 4(x_2 + x_4 + x_6 + \ldots + x_{n-2}) - 3(x_0 + x_n)$$

or

$$d'' = n(n+1) - 4(x_2 + x_4 + x_6 + \ldots + x_{n-1}) - (3x_0 + x_n),$$

depending on whether n is even or odd.

I. If the number of poles is odd (n even) let us divide the set of $n+1$ numbers

$$0, 1, 2, \ldots, n$$

into three parts:

set C_1 containing $\frac{n}{2} - 1$ numbers: $0, 1, 2, \ldots, \frac{n-4}{2}$,

set C_2 containing 2 numbers: $\frac{n-2}{2}, \frac{n}{2}$,

set C_3 containing $\frac{n}{2}$ numbers: $\frac{n+2}{2}, \frac{n+4}{2}, \frac{n+6}{2}, \ldots, n$, and let us denote:

an arbitrary permutation of the numbers of set C_1 by

$$x_2, x_4, x_6, \ldots, x_{n-2},$$

an arbitrary permutation of the numbers of set C_2 by

$$x_0, x_n,$$

and an arbitrary permutation of the numbers of set C_3 by

$$x_1, x_3, x_5, \ldots, x_{n-1};$$

then

$$d' = n(n+1) - 4 \cdot \frac{\dfrac{n-4}{2}\left(\dfrac{n-4}{2}+1\right)}{2} - 3\left(\frac{n-2}{2} + \frac{n}{2}\right)$$

$$= \frac{n^2 + 2n - 2}{2}$$

is the length of the longest path that can be walked while hammering the nails. We cover this distance, beginning from one of the poles of the set C_2, in such a way that we pass in turn from poles of set C_1 to poles of set C_3, and back, and we end at the second pole of set C_2. The length of the path does not depend on the order in which we run through the poles of set C_1, and through the poles of set C_3.

II. If the number of poles is even (n odd), let us denote an arbitrary permutation of numbers

$$0, 1, 2, \ldots, \frac{n-3}{2}$$

by

$$x_2, x_4, x_6, \ldots, x_{n-1},$$

an arbitrary permutation of numbers

$$\frac{n+3}{2}, \frac{n+5}{2}, \frac{n+7}{2}, \ldots, n$$

by

$$x_1, x_3, x_5, \ldots, x_{n-2},$$

and let us assume

$$x_0 = \frac{n-1}{2}, \qquad x_n = \frac{n+1}{2}.$$

Then

$$d'' = n(n+1) - 4 \cdot \frac{\frac{n-3}{2}\left(\frac{n-3}{2}+1\right)}{2} -$$

$$- \left(3 \cdot \frac{n-1}{2} + \frac{n+1}{2}\right) = \frac{n^2+2n-1}{2}$$

is the length of the longest path that can be walked over while hammering nails, and this distance can be covered in different ways.

65. We put the wooden block with its edge along the edge of the table, its corner is one corner of the table; we mark on the table with a pencil how far the block reaches; then we move the block as far as the line drawn, and measure with a ruler the distance in the air from the corner of the table to the further upper corner of the nearest face of the block.

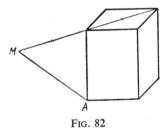

Fig. 82

And now a similar solution: we put the edge of a small board along the diagonal of the upper face of the block, mark that diagonal upon it, pass the board the length of the diagonal (Fig. 82) and measure the distance AM.

66. We draw the trace of the ribbon on the hexahedral sweet box. Then, if we cut the box, and unfold its faces in one plane, the trace becomes a straight line (Fig. 83, on which two faces of the sweet box are repeated). From this figure it is easy to see that

(1) the length of the ribbon is $2\sqrt{(a+c)^2+(b+c)^2}$,

(2) the tangent of the angle at which the ribbon intersects the edges is equal to $\dfrac{a+c}{b+c}$ or $\dfrac{b+c}{a+c}$,

(3) the ribbon can be passed along the box without stretching (in Fig. 83 it would result in parallel shifting),

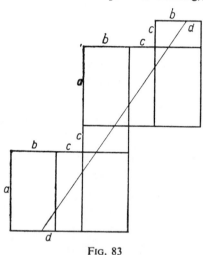

Fig. 83

(4) the length of the ribbon will be the least when on each of the two greatest faces of the box there are two segments of the ribbon; thus if $c < a$ and $c < b$, c being the height of the box, these faces are the bottom and the lid.

67. We limit ourselves to the case where we have a weight of 1 lb.

Let $AC = l$ (Fig. 84) denote the distance between the centre of gravity C of the unloaded club and the point A where the load is attached, and let Q denote the weight of the club. The centre of gravity of the club is found experimentally.

Let us suppose that by loading the club at the point A with a weight equal to 1 lb the club supported at the point B will be in equilibrium. We denote by a the length of the segment AB.

Let D denote the point of balancing when the club is in equilibrium, loaded at the point A with an arbitrary weight of p lb. We denote by x the length of the segment AD.

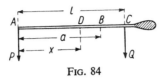

FIG. 84

From the conditions of equilibrium for both cases mentioned above we obtain the equations of moments

$$a = Q(l-a), \quad px = Q(l-x),$$

whence

$$\frac{px}{a} = \frac{l-x}{l-a}. \tag{1}$$

To graduate the scale on the rod of the club we draw the segmen AC, mark on it the point B and from the ends A and C of this segment we lead two parallel rays, m and n, as shown in Fig. 85.

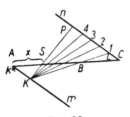

FIG. 85

Then starting from the point C we mark off on the line n segments of length 1, 2, 3, ..., p; through the point L corresponding to a unit and through the point B of AC we draw a segment which intersects the line m at the point K.

From the similarity of the triangles AKB and CLB we obtain for the segment $AK = k$ the expression

$$k = \frac{a}{l-a}, \tag{2}$$

where l is always greater than a.

We shall now project the scale marked off on n on the line AC, choosing the point K as the centre of projection. Let S be the projection of the point P.

From the similarity of triangles AKS and CPS we have, considering relation (2),

$$\frac{p \cdot AS}{a} = \frac{l-AS}{l-a}.$$

Comparing the result obtained with equation (1) we see that $AS = x$; thus S is the point where we must support the club loaded with the weight of p lb.

Thus we see not only how to graduate the club in units of 1 lb but also how to get a finer graduation on the club by providing an auxiliary scale with a sufficiently fine uniform graduation.

From the above reasoning it follows that it was not necessary in the problem to assume constant thickness and uniformity of the rod.

68. As is seen in Fig. 86 we have $AO = \sqrt{OB^2-AB^2}$. As AB is equal to the breadth s of the movable ruler, and the smallest OB is equal to the distance h of the nail from the motionless ruler, the minimum AO is equal to $\sqrt{h^2-s}$.

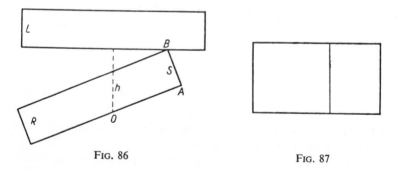

FIG. 86   FIG. 87

69. The primary division of the rectangle into two parts is evident (Fig. 87).

Let us now consider a rectangle divided in the primary way into more than two parts; then every side of the rectangle must be intersected by one of the dividing lines; thus the configuration shown in Fig. 88 for instance is not a primary division, because the vertical line of division cuts from the whole rectangle the left-hand rectangle, the inside of which is already free from additional lines of division, and in the shaded right-hand rectangle we can have an arbitrary system of dividing lines.

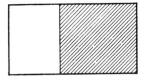

FIG. 88

From this observation it follows that there is no primary division of the rectangle into three parts (Fig. 89), since any division of the rectangle into three parts would leave at least one side untouched, and consequently would not be a primary one. It is also easily seen that there is no primary division of the rectangle into four parts.

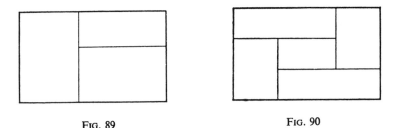

FIG. 89   FIG. 90

The primary division of the rectangle into five parts is, of course, possible (Fig. 90). Similarly, there is a possibility of primary division of the rectangle into 7 or more parts. But the matter is complicated here, since it is possible to give different divisions of the rectangle into the same number of parts. For instance, in Fig. 91 we have two different primary divisions into seven parts, and in Fig. 92 we have four different primary divisions into eight parts.

Let us now proceed to the solution of the problem proper; it refers to primary divisions of the square into a certain number of equal parts.

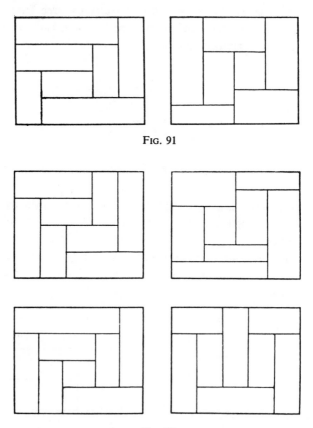

FIG. 91

FIG. 92

Division into 5 equal parts. Let there be given a square with sides equal to 1. Let us suppose that this square has been divided in the primary way into 5 equal parts (Fig. 93).

We limit ourselves to the examination of a symmetric division. Denoting by x the side of one of the little squares, we have $x^2 = 1/5$, whence $x = \sqrt{1/5}$. Then the remaining four small rectangles will have sides $(5-\sqrt{5})/10$ and $(5+\sqrt{5})/10$.

Division into 7 equal parts. Let us consider as before a square with sides equal to 1, and suppose that it has been divided into 7 equal parts according to the scheme in Fig. 94. To denote the sides of the rectangles into which the square is divided let us employ Fig. 94.

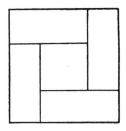

FIG. 93

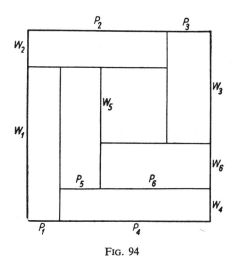

FIG. 94

Let $w_1 = x$. Then $p_1 = 1/7x$ and we have in succession

$$w_2 = 1-x, \qquad p_2 = \frac{1}{7(1-x)},$$

$$p_3 = 1-p_2 = 1-\frac{1}{7(1-x)} = \frac{6-7x}{7(1-x)}, \qquad w_3 = \frac{1-x}{6-7x},$$

$$p_4 = 1-p_1 = 1-\frac{1}{7x} = \frac{7x-1}{7x}, \qquad w_4 = \frac{x}{7x-1},$$

$$w_5 = w_1-w_4 = x-\frac{x}{7x-1} = \frac{x(7x-2)}{7x-1}, \qquad p_5 = \frac{7x-1}{7x(7x-2)},$$

$$p_6 = p_4-p_5 = \frac{7x-1}{7x} - \frac{7x-1}{7x(7x-2)} = \frac{(7x-1)(7x-3)}{7x(7x-2)},$$

$$w_6 = \frac{x(7x-2)}{(7x-1)(7x-3)}.$$

As $w_3+w_4+w_6 = 1$, we get

$$\frac{1-x}{6-7x}+\frac{x}{7x-1}+\frac{x(7x-2)}{(7x-1)(7x-3)} = 1,$$

which, reordered, gives us the equation

$$196x^3-294x^2+128x-15 = 0.$$

One of the roots of this equation is $x = 1/2$. This root is not a solution of the problem, as for $x = 1/2$ we have $w_1 = w_2$, and $p_1 = p_2$, and the division of the square would not be a primary one. The remaining roots of the above equation are $(7+\sqrt{19})/14$ and $(7-\sqrt{19})/14$. According to the conditions of the problem we must have $x > 3/7 = 6/14$, because otherwise p_6 would be negative, so that only the number $(7+\sqrt{19})/14$ satisfies the problem. The sketch given in Fig. 95 corresponds to this solution. It is due to Dr. J. Mikusiński.

We can show that there exists no primary division into 7 equal parts of the sort shown on the right side of Fig. 91. It may be, however, that there exists a primary division into seven equal parts of a certain different kind.

Division into 8 equal parts. Let us now proceed to the primary division of the square into 8 equal parts.

Suppose the possibility of such a division of the unit square, given by the symmetric scheme of Fig. 96; then every partial rectangle has an area equal to 1/8.

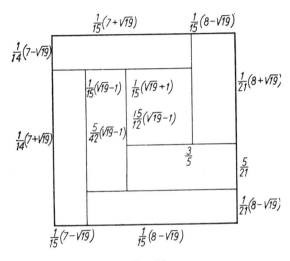

FIG. 95

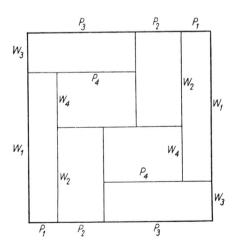

FIG. 96

Because of symmetry, we get $w_2 = 1/2$, and in this case we have $p_2 = 1/4$. Let $w_1 = x$. Then $p_1 = 1/8x$, $w_3 = 1-x$, $p_3 = 1/8(1-x)$, $w_4 = x-1/2$, $p_4 = 1/8(x-1/2)$, and as $p_4 = p_3-p_1$ it follows that

$$\frac{1}{8(x-1/2)} = \frac{1}{8(1-x)} - \frac{1}{8x},$$

whence

$$x(1-x) = x(x-\tfrac{1}{2})-(x-\tfrac{1}{2})(1-x),$$

and, eventually, after simplifications

$$6x^2-6x+1 = 0.$$

From the last equation we obtain $x_{1,2} = (3\pm\sqrt{3})/6$, and, since we must have $x > 1/2$, only the root $x_2 = (3+\sqrt{3})/6$ satisfies the conditions of the problem.

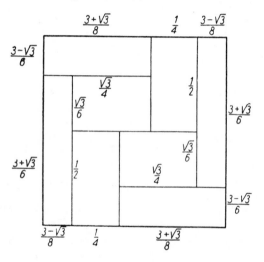

FIG. 97

In Fig. 97 we have a primary division of the square into 8 equal parts, and the dimensions of the partial rectangles are marked there. Besides the division discussed above, one more primary division of the square into 8 equal parts is possible, as shown in Fig. 98; we obtain it in a manner similar to the preceeding one.

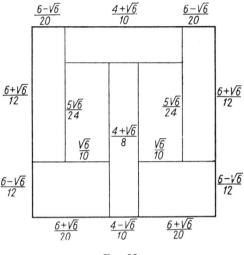

FIG. 98

70. The problem is solved in the following way. We put the small telescope at the point O of the plot of land (Fig. 99), and we measure the inclination of the ground at an arbitrarily chosen direction OC_1. For this purpose we put at the point A_1 a rod pointing in this direction and we measure the distance d_1 and the difference

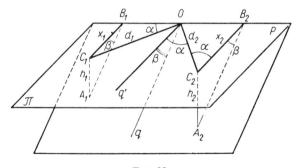

FIG. 99

of levels h_1 between points O and A_1. The inclination of the ground is defined as the ratio $t_1 = h_1/d_1$. Then we rotate the telescope through an angle of $90°$, and measure in the direction OC_2 perpendicular to OC_1 the inclination of the ground, which is expressed by the number $t_2 = h_2/d_2$.

Taking the straight line OC_1 for the x-axis, and OC_2 for the y-axis, we find the vector with components t_1, t_2 on these axes; the direction of this vector is the direction of the inclination of the land, and its length given by $\sqrt{t_1^2 + t_2^2}$ is equal to the inclination of the plot.

Here is the proof of the above result. Let p be contour line passing through the point O; q — the line of inclination; q' — the projection of the line of inclination onto the horizontal plane π; β — the angle of inclination; and let a be the unknown angle which direction OC_2 forms with direction q'.

We have $\tan\beta = h_1/x_1 = h_2/x_2$, whence $x_1/x_2 = h_1/h_2$. Moreover, we have $x_1/d_1 = \sin a$, $x_2/d_2 = \cos a$, whence $\tan a = x_1 d_2/d_1 x_2 = h_1 d_2 = d_1 h_2 = t_1/t_2$; thus the above method of finding the direction of inclination is correct.

Finally we compute the inclination of the land:

$$\tan\beta = h_2/x_2 = h_2/d_2\cos a = (h_2/d_2)\sqrt{1+\tan^2 a} = \sqrt{t_1^2 + t_2^2} = t.$$

71. Let us suppose that the town M is connected by the segments MA, MB, MC, MD, ... with towns A, B, C, D, ... (Fig. 100).

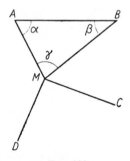

FIG. 100

Let us connect by a segment the point A with B. We have $AM < AB$ and $BM < AB$, since otherwise instead of at least one of the connections AM and BM the connection AB would appear.

From the inequalities

$$AM < AB \quad \text{and} \quad BM < AB$$

we obtain

$$\gamma > \alpha \quad \text{and} \quad \gamma > \beta;$$

adding these inequalities to the identity $\gamma = \gamma$, we obtain $3\gamma > \alpha+\beta+\gamma$; thus $3\gamma > 180°$, and $\gamma > 60°$.

The inequality $\beta > 60°$ shows immediately that the point M can be a common vertex for at most five triangles, which was to be proved.

72. There can be three kinds of networks, as shown in Fig. 101.

FIG. 101

1. In the first case each of the towns can be a node in which four lines are joined; then there will be 5 networks of this kind.

2. In the second case, a town forming a node at which there three lines joined, it can be connected with other 3 towns in 4 ways (the number of combinations of 4 elements taken 3 at the time), and the fifth town in each of these cases can be connected with each of the 3 towns. Then there are $4.3 = 12$ possible connections. And since each of these 5 towns can be a node, then there will be $5.12 = 60$ different networks of the kind discussed.

3. In the third case, in the network joining the towns we arbitrarily change the successive order of the towns. Then we shall obtain as many networks as there are permutations of 5 elements, that is $5! = 120$. But every permutation denotes the same network as the permutation in the opposite order of the elements, whence there will be $120/2 = 60$ different networks.

73. 1. From all triangles with constant bases and constant heights, the one with the shortest perimeter is an isosceles triangle. To prove this it is sufficient to note (Fig. 102) that the point P lying

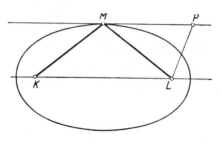

FIG. 102

on the tangent to the ellipse at the point M lies outside this ellipse; thus

$$KP + PL > KM + ML.$$

2. From the above it follows that if points K, L, M are vertices of an isosceles triangle with bases $KL = 2p$ and heights $MO = h$ (Fig. 103), then the network $KS + LS + MS$ is shorter than the network $KP + LP + MP$.

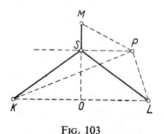

FIG. 103

3. Writing $OS = x$ (Fig. 103) and denoting by m the length of the network $KS + LS + MS$, we have

$$2\sqrt{p^2 + x^2} + h - x = m,$$

whence

$$3x^2 - 2(m-h)x + 4p^2 - (m-h)^2 = 0;$$

thus from the condition of the real character of the solution we obtain

$$(m^2 - h)^2 - 3p \geqq 0.$$

This condition leads to the restriction

$$m \geqq h + p\sqrt{3}$$

for m. Thus m assumes its smallest value for

$$x = \frac{m - h}{3} = \frac{p\sqrt{3}}{3};$$

this happens for $\sphericalangle OKS = 30°$.

4. In the case of connecting the towns A, B, C, D with railways without nodes, the shortest connection would be that shown in Fig. 104 (or one of three other connections which can be obtained

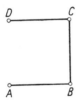

FIG. 104

from this one by rotation through an angle of 90°, 180° and 270° about the centre of the square). The length of such a connection is 300 km.

5. Let us suppose that the towns A, B, C, D are connected with a network containing one node S. This node must be connected with at least 3 of the towns A, B, C, D; otherwise by straightening the path leading through the node we would obtain a shorter network.

Let the node S be connected with towns A, B, C (Fig. 105). Then the town D can be connected either with one of the towns A and C or with the node S.

In the first case, according to what has been proved in paragraphs 2 and 3 ($p = h = 50\sqrt{2}$ km), the length of the shortest network would be (in km)

$$100 + 50\sqrt{2}(1 + \sqrt{3}) \approx 293 \cdot 2.$$

In the second case, according to what has been proved in paragraph 1, the network will be the shortest if the point S lies on the line of symmetry of the side AB of the square, and on the line

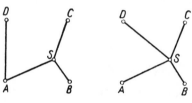

Fig. 105

of symmetry of the side BC of the square, that is, if it lies at the centre of the square; then the total length of the network will be (in km)

$$2 \cdot 100\sqrt{2} \approx 282 \cdot 8.$$

6. Let us suppose that the towns A, B, C, D are connected by a network containing two nodes, S_1 and S_2. The node S_1 has to be connected with at least 3 of the points A, B, C, D, S_2; otherwise a straightening of the path leading through the node would give a shorter network.

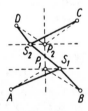

Fig. 106

Let the node S_1 be connected with towns A and B and with the node S_2. Then the node S_2 will be connected with towns C and D and the network will have the configuration shown in Fig. 106.

If the points S_1 and S_2 do not lie on the line of symmetry of the side AB of the square, then, as follows from what was proved in the first part, the network with the nodes S_1 and S_2 is longer than the network with nodes P_1 and P_2, and the latter network is shortest when the points P_1 and P_2 lie symmetrically with respect to the centre of the square, so that $\sphericalangle P_1 AB = \sphericalangle P_2 CD = 30°$. The length (in km) of the network is (see paragraph 3 assuming $p = h = 50$)

$$100(1+\sqrt{3}) \approx 273 \cdot 2.$$

7. An increase in the number of nodes above 2 extends the network. Thus, the shortest network with which towns A, B, C, D can be connected according to the conditions of the problem is (in km)

$$100(1+\sqrt{3}) \approx 273 \cdot 2.$$

It can be realized in two ways, as illustrated in Fig. 107.

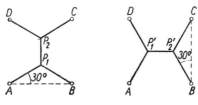

FIG. 107

74. The shortest path on the globe is an arc of the great circle, that is, the circle which is formed when the sphere is cut by a plane passing through the centre of the globe.

As the airplane starting in Oslo disappeared from the eyes of the observers at the airport at the point of the horizon lying exactly to the West, the airplane was flying along the great circle crossing perpendicularly the meridian at Oslo ($a = 10°43'$, Eastern longitude).

In Fig. 108 let S and O denote, respectively, the centre of the globe and the point at which Oslo is situated. And let Q, G and O' denote, respectively, the points of intersection with the equator of the great circle, along which the jet plane was flying, of the Greenwich meridian, and of the meridian passing through Oslo.

It is easy to see that the angle QSO' is a right angle, whence the angle $GSQ = 90° - \sphericalangle GSO' = 90° - 10°43' = 79°17'$; thus the point at which the jet is to land lies on the equator $79°17'$ west of Greenwich (the point of such coordinates lies in the province

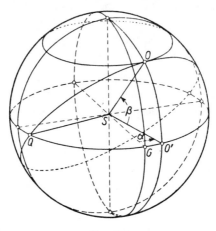

Fig. 108

Pichincha, 120 km west of Quito, the capital of Equador) and as the arc OQ is equal to $1/4$ of the great circle, the path of the flight is about 10000 km.

The plane OSQ is inclined against the plane of the equator at an angle $\beta = 59°55'$; the observers waiting in Equador for the appearance of the plane have to look èxactly in a direction forming an angle β against the East.

75. During a full eclipse of the Sun it is observed that the apparent sizes of the Moon and the Sun are equal (approximately). Therefore it follows that the apparent diameters of the Sun and the Moon are equal (approximately). Therefore we see that the diameter of the Sun is 387 times the diameter of the Moon, and the ratio of volumes is $387^3 = 58.10^6$.

76. The shortest day in Wrocław (that is the period between sunrise and sunset on the same day) is the day of which the sun reaches its lowest position (Tropic of Capricorn); this happens every year around December 23.

To find the length of this day we must know the latitude φ of Wrocław ($\varphi = 51°07'$) and the angle δ made by the plane of the equator with the ecliptic ($\delta = 23°27'$; the values of both angles are approximate).

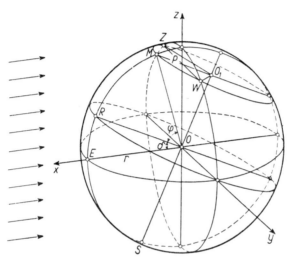

Fig 109

Let (Fig. 109) letters O and M denote, respectively, the centre of the globe and the location of Wrocław on the globe, and let the straight line NS denote the axis of rotation of the globe. Let the plane of the ecliptic be the plane xOy; the plane of the equator, passing through the point R, forms with it the angle δ. The plane OWZ, perpendicular to the plane of the ecliptic and to the direction Ox of falling Sun's-rays, separates on the globe the illuminated part from the dark one. Wrocław passes at sunrise through the point Z, runs along the arc ZMW and is at sunset at the point W.

The length of the shortest day in Wrocław is evidently proportional to the length of the arc ZMW of the parallel passing through Wrocław, or — shortly — to the magnitude of the angle $2\beta = \sphericalangle ZO_1W$.

Looking at the rectangular triangles OO_1M, OO_1P and O_1PW of Fig. 109 we shall easily find

$$\cos\beta = \frac{O_1P}{O_1W} = \tan\varphi\tan\delta,$$

whence $\beta = 57°27'35''$.

For the length t of the shortest day in Wrocław we have the proportion

$$t:T = \beta:180° \quad (T = 24 \text{ hours}):$$

hence $t = 7$ hours 39 minutes and 40 seconds.

It is worth while to stress that the true length of the shortest day in Wrocław is greater than that which we have just computed by about a quarter of an hour. This is due to the breaking up of the Sun's rays by the layers of the atmosphere surrounding the globe ("astronomical refraction").

77. A square which is adjacent to a white square is a black one, and *vice versa*. On the chessboard with an odd number of squares the number of white squares is not equal to the number of black ones; thus the movement of the pawn required in the problem is impossible.

78. The answer to the question is in the negative. No pawn will change its place, and it does not matter whether by neighbouring squares we mean only two squares adjacent along the side, of whether we mean the squares which have only one common vertex. In the latter case it is indispensable to assume that our quadratic chessboard has more than four squares. The proof can be carried out jointly for both cases as follows: Let us denote by (i, k) the square located in the i-th row and k-th column of the chessboard with n^2 squares $(n > 2)$ and let us denote by $F(i, k)$ the square on which we put the pawn standing formerly on the square (i, k). Assuming that $F(1, 1), = (1, 1)$, $F(n, 1) = (n, 1)$, and that the transformation $F(i, k)$ preserves the neighbourhood of squares, we have to prove that $F(i, k) = (i, k)$ for $i = 1, 2, ..., n$ and $k = 1, 2, ..., n$.

We immediately see that the number of squares which are neighbours of a certain defined square of the chessboard is smaller for a corner square than for the remaining border squares, and that for border squares it is smaller than for interior ones. Since in its new position no pawn can have fewer neighbours than it had before, then if the square (i, k) is interior, the square $F(i, k)$ is interior also, or both are corner squares, or finally both are bordering and not corner squares.

$(1, 1), ..., (2, 1), ..., (n, 1)$ is a sequence of successive neighbouring border squares. The sequence of squares $F(1, 1), F(2, 1), ..., F(n, 1)$ must have the same property. Thus $F(1, 1) = (1, 1)$, but the square $(1, 1)$ (assuming that $n > 2$) adjoins only two outside border squares $(1, 2)$ and $(2, 1)$. Thus the function $F(2, 1)$ can have only one of these values. Let us consider these two cases:

(a) $F(2, 1) = (1, 2)$. In this case the square $F(3, 1)$, being a border square adjoining the square $F(2, 1)$, i.e. square $(1, 2)$ and different from $(1, 1)$, would have to coincide with the square $(1, 3)$; similarly we would have $F(4, 1) = (1, 4)$ etc. Finally the corner square $F(n, 1)$ would be the square $(1, n)$, contrary to the assumption that $F(n, 1) = (n, 1)$. Thus case (a) cannot occur and it is the second case (b) $F(2, 1) = (2, 1)$ which occurs.

(b) $F(2, 1) = (2, 1)$. In this case, reasoning as in (a), we obtain successively $F(3, 1) = (3, 1)$, $F(4, 1) = (4, 1)$., ..., $F(n, 1) = (n, 1)$ which shows that the pawns of the first column remain in their places.

The equations $F(i, 1) = (i, 1)$ where $i = 1, 2, ..., n$ imply equations $F(i, 2) = (i, 2)$ for $i = 1, 2, ..., n$. In fact, the square $F(1, 2)$ is a border square adjoining the square $(1, 1)$ and different from the square $(2, 1)$, and thus $F(1, 2) = (1, 2)$; the square $F(2, 2)$ is a neighbour of the squares $F(2, 1)$ and $F(1, 2)$, that is of the squares $(2, 1)$ and $(1, 2)$, and is different from $(1, 1)$; thus $F(2, 2) = (2, 2)$ etc.

In the same way we verify the equation $F(i, k) = (i, k)$ for the next successive column of the chessboard.

Generalization. The result obtained above allows us to draw a certain general conclusion.

Let us consider function F which assigns to each square of the chessboard a certain defined square of the chessboard in such a way that to two different squares correspond different squares, and that to neighbouring squares correspond also neighbouring squares.

Such functions (transformations) are for instance: identity (it is the transformation which leaves each square in the same position), the symmetries with respect to each of the 4 axes of symmetry of the chessboard, as well as rotations of 90°, 180° and 270° about the centre of the chessboard.

We shall show that every transformation F of the required kind has to be one of the eight transformations mentioned above.

In fact let F be such a transformation. Since, as we know, corner squares correspond to corner squares, then only the following cases can occur:

I. Transformation F leaves the square $(1, 1)$ in its place. In this case one of two possibilities occurs:

1. The square $(n, 1)$ remains in its place. Then the transformation F is, as shown before, an identity.

2. The square $(n, 1)$ passes into another corner square. This square can be only $(1, n)$, which is proved by reasoning as before, i.e. by considering the sequence $F(1, 1)$, $F(2, 1)$, ..., $F(n, 1)$. Let us carry out after the transformation F a symmetry S_1 with respect to the main diagonal passing through $(1, 1)$ and (n, n). The product ([1]) $F.S_1$ is a transformation of the required kind, which leaves each of the squares $(1, 1)$ and $(n, 1)$ in their respective places; thus it is the identity transformation I:

$$F.S_1 = I.$$

Multiplying this equation on the right by S_1 we obtain $(F.S_1).S_1 = I.S_1$ or $F.S_1^2 = S_1$, and as $S_1^2 = I$, then $F = S_1$; in this case the transformation F is a symmetry with respect to the main diagonal of the chessboard.

II. The square $(1, 1)$ passes, by transformation F, into the square (n, n). Let S_2 denote the symmetry with respect to the second diagonal of the chessboard; it transforms the square (n, n) into the square $(1, 1)$. Thus the transformation $F.S_2$ leaves the square $(1, 1)$ at the same place; consequently, according to I, two possibilities occur: $F.S_2$ is either an identity or a symmetry.

3. The product $F.S_2$ is an identity, that is

$$F.S_2 = I.$$

([1]) Let A, B denote the transformation of a certain set into itself. $A.B$ is the transformation resulting from submitting this set first to the transformation A, then to the transformation B. If, for instance, A is an axis of symmetry on the plane, then $A.A$, that is A^2, is an identity transformation; if A and B are symmetries with the respect to the intersecting axes a and b, then $A.B$ is a rotation about the point of intersection of axis a and b, by an angle twice as great as the angle between the axes. If I is an identity transformation, and A an arbitrary transformation, then $A.I = I.A = A$. The product of transformations obeys the associative law $(A.B).C = A.(B.C)$.

Multiplying this equation on the right by S_2 we obtain

$$F = S_2,$$

thus F is a symmetry with respect to the diagonal of the chessboard passing through the squares $(1, n)$ and $(n, 1)$.

4. The product $F.S_2$ is the symmetry S_1:

$$F.S_2 = S_1.$$

Multiplying this equation on the right by S_2 we obtain

$$F = S_1.S_2.$$

Thus the transformation F is in this case a rotation by $180°$ about the centre of the chessboard, or the centre of symmetry with respect to this point.

III. As a result of the transformation F the square $(1, 1)$ passes into the square $(n, 1)$. In this case the transformation $F.S_3$, where S_3 denotes the symmetry with respect to the horizontal axis of symmetry of the chessboard, leaves the square $(1, 1)$ at the same place and we again have two possibilities:

5. $F.S_3 = I$, whence $F = S_3$,

6. $F.S_3 = S_1$, whence multiplying the right side by S_3, we obtain

$$F = S_1.S_3,$$

which shows that the transformation F is a rotation through $90°$ about the centre of the chessboard.

IV. The transformation F takes the square $(1, 1)$ into the square $(1, n)$. Denoting by S_4 the symmetry with respect to the perpendicular axis of symmetry of the chessboard, we obtain, as in case III, two possibilities:

7. $F = S_4$

or

8. $F = S_1.S_4$; this shows that the transformation is a rotation by $270°$ about the centre of the chessboard.

Thus we have proved that any transformation of the set of chessboard squares into itself, preserving the contiguity of squares, is one of the eight transformations:

$$I, S_1, S_2, S_3, S_4, S_1 S_2, S_1 S_3, S_1 S_4.$$

These transformations form a group.

Returning to the problem of pawns, on the basis of this theorem, we can formulate a more general result:

If in rearranging the pawns we are to leave in its former place at least one pawn not standing on any axis of symmetry of the chessboard, then each pawn must retain its former position.

79. The problem is solved if we give a method of placing rooks on the chessboard which is consistent with conditions (1), (2), (3), (4). For simplicity, let us introduce certain terms. A column is called *free* if there is no rook standing in it and if it contains at least one white square not attacked horizontally by any rook standing on the board. We call a column *occupied* if it has a white square with a rook on it, this rook not being menaced horizontally by any other rook standing on the chessboard. We call a row *free* if there is no rook in it.

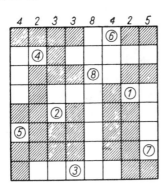

FIG. 110

On the empty chessboard we have as many free columns as there are free rows. Let us first write over each column the number of white squares in this column and then proceed to place the rooks successively. We always place a rook in a free column over which there is a number not greater than the number over any of the remaining free columns. If we succeeded in placing the rooks in this way in all the columns then the problem would be solved, because the conditions (1), (2), (3), (4) would be satisfied. An example of such an arrangement is shown in Fig. 110. The small circles on the white squares denote rooks, and the numbers inside the circles denote the order of their placement. Evidently, it is not the unique solution for this chessboard.

However, the procedure described above may encounter certain obstacles:

(a) On placing one of the rooks we may find that all the white squares of one or more columns in which a rook has not been placed yet are already horizontally attacked by the rooks placed before. We cannot place a rook in such a column without violating condition (3) of the problem, and we must do it to satisfy condition (4).

Then we try to move vertically one of the already placed rooks to another white square of the same column. If we succeed we can continue to place rooks. Such an example is shown in Fig. 111.

We see that after placing 4 rooks on the first chessboard according to the method given above, all of the white squares of column *d* and *e* are already horizontally attacked. Now we look for a column in which there is already a rook placed but there are still free white squares not attacked horizontally by any of the remaining rooks. Such a column is, for example, column *g*. Then we move rook 2 onto square *g*5 (the second chessboard in Fig. 111). Now, we can put the fifth rook on square *d*4 in column *d*. Analogously, we move rook 3 to the square *b*8 and then we put the sixth rook in column *e*. In this way we can remove the obstacle, and continue to place the rooks in the way described (last chessboard in Fig. 111).

It is possible that we encounter only obstacles that can be removed by the trick described above. Then we can place the rooks on the chessboard without difficulty according to conditions (1), (2), (3) and (4).

(b) It can occur, however, that in one or more columns all white squares are attacked horizontally, and moreover, in the columns in which we have already placed rooks there are no free white squares. We shall say that such columns are *blocked*. In blocked columns we can place no rook, as it would be in contradiction to condition (3). Neither can we move the rooks already placed, because in the columns in question there are no free white squares.

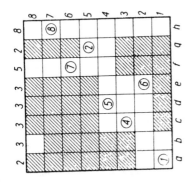

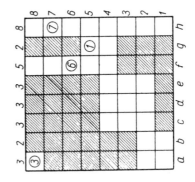

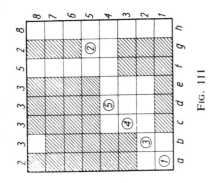

FIG. 111

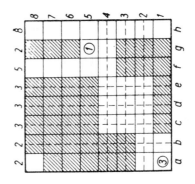

FIG. 112

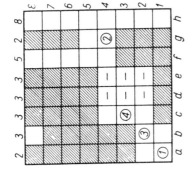

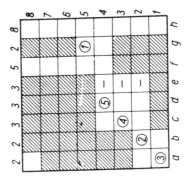

We shall show an example how to eliminate such an obstacle. On the first chessboard in Fig. 112 we see that column e is blocked: all the white squares of this column are attacked by rooks 2, 4, 5, and moreover, there are no free white squares in any of the columns occupied by these rooks. Now we cease altogether to place rooks in the blocked columns and the columns and rows from which we removed the blocking rooks. In our example, these are columns b, c, d and e, and rows 2, 3 and 4. It is easy to see that the number of eliminated rows is always less than the number of eliminated columns, and that the squares in the eliminated columns and in the free rows are black. Thus we do not have to worry that by additional placing of rooks there will appear in the eliminated columns white squares attacked horizontally by newly placed rooks.

It can happen that in the eliminated columns there remain white squares attacked horizontally by rooks already placed and not removed by eliminating the blockade. According to condition (4) these squares have to be attacked vertically by a certain rook. However, we shall not place any more rooks in the eliminated column — we shall call this situation a secondary blockade. In our example the square b_1 (Fig. 112) of the eliminated column b is attacked horizontally by rook 1. We have two ways of eliminating such a secondary blockade. At first we verify whether in the column in which the blocking rook is located there are still free white squares in non-eliminated free rows. If that is so, we try to move the blocking rook to a white square in the column, and then we continue placing the rooks. If, however, this column also has no more free white squares, then, in addition, we eliminate the row and column in which the blocking rook is located. Of course, we remove this rook. In our example (Fig. 112) rook 3 can be moved onto square $a8$, and in this way we eliminate the secondary blockade. If square $a8$ is black, then we shall also eliminate column a and row 1. In both cases further placing of rooks in our example does not encounter any obstacle.

Having eliminated the secondary blockades and rejected suitable rows and columns, we continue to place the rooks. We can encounter additional blockades, original and secondary ones, which we eliminate analogously.

It can be shown that by placing the rooks in the way described above, we shall place on the chessboard at least one rook according to the condition of the problem. Let us first see that, by always placing the rooks in a free column in which there are at most as many white squares as there are free columns, we shall place the last rook in a column in which there are only white squares. We shall always be able to place the rook in this column, independent of whether blockades of both types appeared before or not.

If the blockades do not appear, then there will be $n-1$ rooks on the chessboard, each of them in a different column and a different row. Thus only one free row and one free column will remain. As the whole column is white, we place the rook in the free row. Then there will be n rooks on the chessboard, each of them in a different row and in a different column, which do not threaten one another. Thus all the squares not occupied by rooks are attacked vertically and horizontally. Such an arrangement satisfies the conditions of the problem.

If, on the other hand, a blockade appears, then eliminating all the rejected rows and columns and reassembling the remaining squares we should obtain a rectangular chessboard in which there are more rows than columns. By placing the last rook in the white column we cannot make a blockade of any column, as there are already rooks placed in each column, and thus each square not eliminated, and at the same time not occupied by a rook, is attacked vertically.

Neither can it occur that the last white column gets blocked. This follows from the fact that previously we could place at most $n-1$ rooks which attack horizontally $n-1$ squares of the white columns; consequently we can place the last rook in the n-th square of the column, which is not attacked.

It is worth-while to remark that the given solution is valid for all rectangular chessboards which satisfy the assumptions of the problem and in which the number of columns is less than the number of rows.

80. We take advantage of the following property: a tangent to the ellipse at an arbitrary point forms equal angles with the seg-

ments joining this point with the foci of the ellipse. It means that
the billiard ball struck from the focus F_1 hits focus F_2 (in Fig. 113
the track of the billiard ball is marked by a dotted line). Therefore
the assumption that the problem can be solved leads, as the diagram
shows, to a contradiction of the law of reflection, since it demands
that the angle of incidence be greater than the angle of reflection.

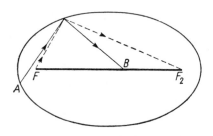

FIG. 113

81. There is a total of 25 pupils in the class. As 6 pupils have grades
D, E or F in mathematics, there are 19 pupils with at least grade C,
and there are at most the same number of sportsmen. The "points",
counting 1 point for any sport practised by one pupil, are $17+13+$
$+8 = 38$ in the class, and as no pupil practises all 3 sports, then
considering the above remark there are precisely 19 sportsmen
and each of them trains for 2 sports. Now it is easy to answer the
questions put in the problem.

(a) No pupil in the class has grade A in mathematics.

(b) Of the 19 sportsmen 17 can ride a bicycle. Consequently
there are only two pupils who can both swim and ski. Thus there
are two swimmers who can ski.

82. On the first day the runners finished the race in the order
A, B, C. On the second day the order was B, C, A, and on the
third day C, A, B. The result of these races is that A proved to
be faster than B in two races, and B surpassed him only in one; B
proved faster than C in two out of three runs, and C also proved
to be faster than A in two out of three runs. Is it possible that such
results appear in three out of four runs, that is, in 75%, or more,
of the runs?

83. It is impossible that all members of the club should form one class, that is, that they should have an equal number of victories, since the number of games (45) in not divisible by 10.

It is also impossible to split the competitors into 9 classes. In fact if there were 9 classes, then there would be 2 players in one of them and one player in each of the remaining 8 classes. This is contrary to the conditions of the problem.

In fact, let us give to each player one point for each game won. The nine players representing nine classes could not have either $0+1+2+...+8 = 36$ or $1+2+3+...+9 = 45$ points, because then the tenth player would have 9 or 0 points and that would make 10 classes again.

But neither can the same 9 players have

$$0+1+...+i+(i+2)+...+9 = 36+8-i$$

points, because the ninth player would have to have $i+1$ points, and again there would be 10 classes.

Yet it is possible to split the competitors into 10, 8, 7, 6, 5, 4, 3, and 2 classes.

We give the examples of division into 7 classes and into 3 classes. The players are denoted by letters. The digit 1 on the intersection of the i-th row with the k-th column denotes a victory of the player in row i over the player in column k, and 0 denotes a defeat.

Splitting into seven classes

	A	B	C	D	E	F	G	H	I	J	
A		1	1	1	1	1	1	1	1	1	
B	0		1	1	1	1	1	1	1	1	
C	0	0		1	1	1	1	1	1	1	
D	0	0	0		1	1	1	1	1	1	
E	0	0	0	0		0	1	1	1	1	(i)
F	0	0	0	0	1		0	1	0	1	
G	0	0	0	0	0	1		0	1	0	
H	0	0	0	0	0	0	1		0	1	
I	0	0	0	0	0	1	0	1		0	
J	0	0	0	0	0	0	0	1	0		

(k)

Splitting into 3 classes

	A	B	C	D	E	F	G	H	I	J
A		1	0	1	1	1	1	1	1	0
B	0		1	1	1	1	1	1	1	0
C	1	0		0	1	1	1	1	1	1
D	0	0	1		1	1	1	1	1	1
E	0	0	0	0		1	0	0	1	1
F	0	0	0	0	0		0	1	1	1
G	0	0	0	0	1	1		0	0	1
H	0	0	0	0	1	0	1		0	1
I	0	0	0	0	0	0	1	1		1
J	1	1	0	0	0	0	0	0	0	

What examples can be given for splitting the competitors into 10, 8, 6, 5, 4 and 2 classes?

84. Let us prove by mathematical induction that a certain team is always a leader. Let there be n teams in the league. We gather the captains of these teams in a room, and order one of them (let us call him K) to leave the room together with the captains of all the teams defeated directly by team K. Then n' captains will remain in the room; $n' < n$. If the theorem is true for all n' less than n, then in the room we already have the leader-captain with respect to all those remaining in the room. As he has also remained in the room, he is a direct winner over K, and indirect winner over those who left the room — thus he is leader of the league of all n teams, which is what we wanted to show.

The proof of (2) is just as easy. Let us call all the teams D_1, $D_2, ..., D_n$. Let D_1 be the team which had the greatest number of direct victories. Suppose, for instance, that the teams D_2, D_3, $..., D_m$ were beaten directly by D_1 and no other team suffered such a direct defeat from team D_1. It is to be proved that D_1 defeats indirectly teams $D_{m+1}, ..., D_n$. Let us suppose the contrary, which means that, for example, there is a team D_{m+1} which is defeated by none of the teams D_2 and D_3 ... and D_m; it follows that it has directly defeated all those teams. As D_1 does not defeat (directly) D_{m+1}, then D_{m+1} defeats D_1; thus D_{m+1} defeats directly more teams than D_1 — contradicting the assumption made about D_1. This contradiction proves (2).

85. The question must be understood in the following way: In the cup system it can happen that the player who deserves the second prize because he is stronger than all the other players, with the exception of the champion, will be conquered by the champion before the final round, that is, in the case of 8 players, in the first or second round; consequently, the finalist will receive the second prize unjustly. This can happen if and only if, in drawing the lots which are to determine the order of 8 players, the player who deserves the title of vice-champion appears in the same group of four players as the champion. Let us denote the champion by C, the vice-champion by V; as in this group in which there is C there are 3 places of which every one can be assigned to V, then there are 7 places for V of which 4 are advantageous and 3 are not. The probability that V will obtain the title of vice-champion is thus $4/7 = 0.5714...$, i.e. more than 57%.

86. In Fig. 114 we mark off on the horizontal axis the path to the post office on the left hand, and the path to the shop on the right. On the perpendicular axis we mark off the time. The straight lines OP_1 and $OO'P_2$ are graphs of the routes of the messengers. If the bicyclist pursues the earlier messenger first, then his route will be described by the straight line $KA_1B_1C_1D_1$; if he pursues

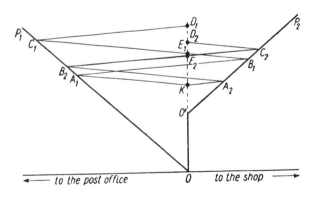

FIG. 114

the later one first, his route will be described by the straight line $KA_2B_2C_2D_2$. We see from the figure that the bicyclist ought to choose the second possibility: this rule applies also to the case when he has only to give money to the messengers, that is, when his route ends at the point E_1 or E_2.

87. Since every dog runs at right angles to the direction of the dog which runs after him, and this dog runs directly after him, then the chasing dog approaches his neighbour with the speed of 10 m/sec and will catch him after 10 sec. Thus the path of each dog is equal to 100 meters. At every moment the four dogs form a square. This square rotates and decreases in area, its sides decrease uniformly at the speed of 10 m/sec. The tracks will intersect at the centre S of the original square. They will be curves (logarithmical spirals). They will not intersect earlier, since, if any of the dogs stepped on the track of another dog, it would mean that the other dog had been earlier in the same spot, which is impossible considering that the four distances of the dogs from S are always equal and steadily diminishing.

88. Let α denote a temporary angle between the direction PQ and the track of ship Q (Fig. 115) and v the speed of ships P and Q

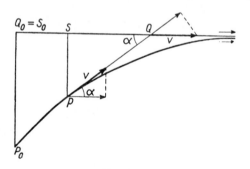

FIG. 115

at the same moment. The mutual approach of the ships depends on the speed v of ship P directed to Q on the component $v \cos \alpha$ of the speed of ship Q — both of the same orientation. Therefore the ships approach each other with a speed equal to $v(1-\cos \alpha)$.

The projection S of point P on the track of ship Q moves along this track with the speed $v - \cos\alpha$, and ship Q escapes with the speed v, thus the distance SQ increases with the speed $v(1-\cos\alpha)$. Since, as we have remarked already, the distance PQ decreases and the distance SQ increases with the same speed, the sum $PQ+SQ$ is constant, and is always equal to 10 miles, as at the initial moment. After an infinite time P will coincide with S, and there will be $PQ+SQ = 2PQ = 10$ miles, and $PQ = 5$ miles.

89. Let P_1, P_2 and O_1, O_2 denote the positions of the two ships, respectively, at the initial moment (when the second ship is sight-

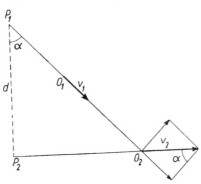

FIG. 116

ed by the first one), and at the moment when their distance is the least. If so, then the first ship was sailing from the point P_1 to O_1 along the straight line P_1O_2 (Fig. 116). Moreover, at the terminal moment the components along O_1O_2 of the speeds of both ships must be equal, that is $v_1 = v_2\sin\alpha$, $\sin\alpha = k$, where the angle α denotes the course of the first ship. As the ships reach the points Q_1, Q_2 at the same time, we get

$$\frac{P_1O_1}{P_2O_2} = \frac{v_1}{v_2} = \sin\alpha.$$

Since $P_2O_2 = d\tan\alpha$, $P_1O_2 = d/\cos\alpha$, the least distance of the ships is equal to

$$O_1O_2 = \frac{d}{\cos\alpha} - \frac{d\sin^2\alpha}{\cos\alpha} = d\cos\alpha = d\sqrt{1-k^2}.$$

90. If k has the same meaning as in the former problem, then the course of the signalling ship is determined by the angle α (Fig. 116) defined by $\sin\alpha = k = v_1/v_2$. If — on the contrary — k denotes the ratio of the speed of the observed ship to the speed of thy signalling ship, then for a course defined again by the equalite $\sin\alpha = k$ the ships will meet at the point O_2 (Fig. 116), because

$$\frac{P_1 O_2}{v_1} = \frac{P_2 O_2}{v_2},$$

i.e.

$$\frac{1}{\cos\alpha . v_1} = \frac{\tan\alpha}{v_2}, \qquad \sin\alpha = \frac{v_2}{v_1} = k.$$

Thus in the second case the course is to be found by the same computation as in the first case.

91. The observing ship is, at the moment mentioned in the problem, at the point S, and the motor boat is at the point M (Fig. 117). The boat moves along sides MN, NW of the square. The observing ship will not reach the boat on the segment MN because she is too slow. Let us denote by a the length of the side of the square, by v the speed of ship (thus the speed of the boat is equal to $3v$).

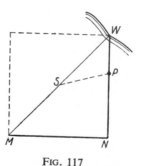

FIG. 117

The point P lying on the segment NW is dangerous for the boat if the observing ship can reach it not later than the boat, that is, if

$$\frac{MN+NP}{3v} \geqq \frac{SP}{v},$$

which implies

$$a + NP \geq 3 . SP, \quad (a + NP)^2 \geq 9(SP)^2.$$

As $(SP)^2 = (NP)^2 + a^2/2 - a . NP$, the above inequality is equivalent to the following inequality for NP:

$$16(NP)^2 - 22a . NP + 7a^2 \leq 0.$$

The inequality is satisfied for $a/2 \leq NP \leq 7a/8$, and thus the dangerous part of the route amounts to 3/16 of the whole; it begins when the boat has passed 3/4 of the way, and terminates when she has passed 15/16 of the way.

92. At the initial moment the boat is at point M and the observing ship at point S. The boat is to reach the coast at point W (Fig. 118).

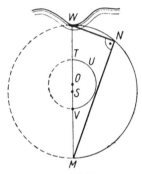

FIG. 118

The points M, S, W lie on the same straight line, $MS = SW$. From the wording of the problem it follows that the boat is to sail along the arms of the right angle MNW, inscribed in the circle with a diameter equal to MW.

The path of the boat is to be the shortest possible, and thus its first segment (MN) should form the least angle with the direction MW.

It is also required that the boat should escape from the observing ship. Therefore its track cannot cross the Apollonian circle TUV, which is the geometrical locus of points P such that the ship and the boat sailing from points S and M respectively with their maximum speed along straight lines meet at P. The centre of this Apollonian circle is the point O, whose position on the segment MW is found from $MT = 3.ST$, $MV = 3.SV$, $OT = OV$. From these equalities it follows that $MO = 9MW/16$, and the radius of the Apollonian circle is equal to $3MW/16$. If the segment MN were tangent to the circle TUV, then the route of the boat would be equal to $(2\sqrt{2}+1)MW/3$, and if MN has no common points with this circle, then the route is longer than $(2\sqrt{2}+1)MW/3$. We have only to verify that, if the segment MN passes sufficiently close to the circle TUV, then the observing ship is also unable to reach the boat on the segment NW. We leave the proof to the reader. The course of the boat satisfying the conditions of the problem is thus found.

93. Such a strange number is number 1, which can be written in three ways: 1, 100%, and $57°17'44''8...$ (radian).

94. With the new tailors' tape it is possible to have an error of 1/2 cm, the same as in measuring with the old one. However, it is not with respect to this point that we need to emphasize the virtues of the new tape. When we measure a certain segment *AB*, then in general the end *B* falls between two marks indicating the unit of measure. Using the new tape (Fig. 119) we attribute to this

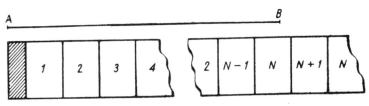

Fig. 119

segment the length of *N* centimeters if *N* is the inscription between the two marks. On the other hand, using the old tape we hesitate as to whether *B* lies nearer to the right or to the left mark. In the exceptional case where, while using the new tape, we find that the end *B* coincides with the mark lying between numbers *N* and *N*+1, we write $AB = N+1/2$ cm.

95. No dictionary contains more than one million words. If, for example, it has 500 pages, we open it on page 250 and find that the last word on it is, say, *narcissus*. The first question will be whether the word to be guessed is located in the dictionary after the word *narcissus*. If the answer is "yes", we open the dictionary on page 375, if it is "no", we open the dictionary on page 125, and we again look for the last word on the open page, etc. If such a division gives a fraction of the page (for example page 62·5), we round it off to the next higher number (63). After 9 such questions the word to be found is restricted to one page. If there are 2 columns on this page, then the tenth question will designate the column. In a column there are at most 64 words, and consequently the sixteenth question will restrict us to one word, which we announce as the solution. The dictionary which we give as an example has no more than $128 \cdot 500 = 64000$ words. If there were 16 times more, or more than a million, then we would require 4 additional questions ($2^4 = 16$), that is altogether 20 questions ($2^{20} = 1024000$). Such a procedure observes the first rule of the quiz, namely that the question must be answerable by "yes" or "no", and profits by the second rule, namely that such categorical answers are obligatory.

96. Each student lending somebody money denotes the sum by a positive number, and borrowing money from somebody denotes the amount of money by a negative number. The sum of all noted numbers gives at the end of the year a positive or negative balance for the given student, but the sum of all the balances is zero.

The book-keeping of Dr. Abracadabrus suffices to pay the debts, and at most 6 payments are necessary. In fact, let us suppose that the "smallest" debtor pays the whole sum to the "largest" creditor: then after the first payment the whole group of debtors and creditors will be equal to at most six, and after the fifth payment, based on the same principle, the number of debtors and creditors will be at most 2. In view of this, all the debts will be paid after at most 6 payments. It can happen, of course, that after a certain succesive payment one of the creditors will become a debtor, but this fact will not change the number of debtors and creditors.

97. Such a social set will be found if we place each of the 12 persons unknown to one another on one of the faces of a regular dodecahedron and require each person to become acquainted with his neighbours, that is, with the inhabitants of neighbouring pentagons; we say that two pentagons are neighbours if they have one common side.

Condition (f) is satisfied because the persons placed on opposite faces have no common neighbours.

The social set of Dr. Abracadabrus also inhabited a regular dodecahedron, but in this case the contrary rule was obligatory, namely to ignore neighbours and become acquainted with all non-neighbours.

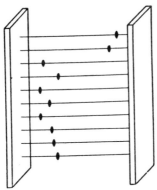

FIG. 120

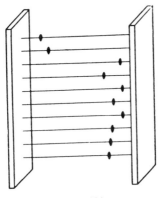

FIG. 121

98. Doctor Abracadabrus is right. Let us suppose that it is otherwise, i.e. that at a certain moment there are only two or less than two beads on the right side of the abacus (Fig. 120). After the time each of the beads requires to cover the whole wire, the situation will be symmetrical to the former one (Fig. 121). In the left half of the abacus there will be two or less than two beads, and in the right half 8 or more beads, which contradicts the assumption made in the problem.

99. The decision of Dr. Abracadabrus is correct, independent of where he lives. In fact, Fig. 122 shows how to cover all streets

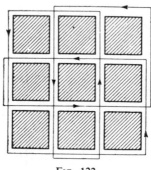

FIG. 122

of the town and return to an arbitrary starting point. The length of the shortest of such tracks exceeds by 16·7% the total length of the town streets.

100. The five cities, Châlons, Vitry, Chaumont, St. Quentin and Reims, form a closed pentagon; one of its sides (side Chaumont-St. Quentin) is the sum of the remaining four sides because 236 = 86+40+30+80. This is possible only in the case where the vertices of the pentagon lie on the same straight line. The towns lie in the following order: St. Quentin, Reims, Châlons, Vitry and Chaumont. Consequently the distance Reims-Chaumont is 40+30+80 = 150 km.

A CATALOG OF SELECTED

DOVER BOOKS

IN ALL FIELDS OF INTEREST

A CATALOG OF SELECTED DOVER
BOOKS IN ALL FIELDS OF INTEREST

CONCERNING THE SPIRITUAL IN ART, Wassily Kandinsky. Pioneering work by father of abstract art. Thoughts on color theory, nature of art. Analysis of earlier masters. 12 illustrations. 80pp. of text. 5⅜ x 8½.   23411-8 Pa. $4.95

ANIMALS: 1,419 Copyright-Free Illustrations of Mammals, Birds, Fish, Insects, etc., Jim Harter (ed.). Clear wood engravings present, in extremely lifelike poses, over 1,000 species of animals. One of the most extensive pictorial sourcebooks of its kind. Captions. Index. 284pp. 9 x 12.   23766-4 Pa. $14.95

CELTIC ART: The Methods of Construction, George Bain. Simple geometric techniques for making Celtic interlacements, spirals, Kells-type initials, animals, humans, etc. Over 500 illustrations. 160pp. 9 x 12. (USO)   22923-8 Pa. $9.95

AN ATLAS OF ANATOMY FOR ARTISTS, Fritz Schider. Most thorough reference work on art anatomy in the world. Hundreds of illustrations, including selections from works by Vesalius, Leonardo, Goya, Ingres, Michelangelo, others. 593 illustrations. 192pp. 7⅛ x 10¼.   20241-0 Pa. $9.95

CELTIC HAND STROKE-BY-STROKE (Irish Half-Uncial from "The Book of Kells"): An Arthur Baker Calligraphy Manual, Arthur Baker. Complete guide to creating each letter of the alphabet in distinctive Celtic manner. Covers hand position, strokes, pens, inks, paper, more. Illustrated. 48pp. 8¼ x 11.   24336-2 Pa. $3.95

EASY ORIGAMI, John Montroll. Charming collection of 32 projects (hat, cup, pelican, piano, swan, many more) specially designed for the novice origami hobbyist. Clearly illustrated easy-to-follow instructions insure that even beginning papercrafters will achieve successful results. 48pp. 8¼ x 11.   27298-2 Pa. $3.50

THE COMPLETE BOOK OF BIRDHOUSE CONSTRUCTION FOR WOODWORKERS, Scott D. Campbell. Detailed instructions, illustrations, tables. Also data on bird habitat and instinct patterns. Bibliography. 3 tables. 63 illustrations in 15 figures. 48pp. 5¼ x 8½.   24407-5 Pa. $2.50

BLOOMINGDALE'S ILLUSTRATED 1886 CATALOG: Fashions, Dry Goods and Housewares, Bloomingdale Brothers. Famed merchants' extremely rare catalog depicting about 1,700 products: clothing, housewares, firearms, dry goods, jewelry, more. Invaluable for dating, identifying vintage items. Also, copyright-free graphics for artists, designers. Co-published with Henry Ford Museum & Greenfield Village. 160pp. 8¼ x 11.   25780-0 Pa. $10.95

HISTORIC COSTUME IN PICTURES, Braun & Schneider. Over 1,450 costumed figures in clearly detailed engravings–from dawn of civilization to end of 19th century. Captions. Many folk costumes. 256pp. 8⅜ x 11¾.   23150-X Pa. $12.95

CATALOG OF DOVER BOOKS

EARLY NINETEENTH-CENTURY CRAFTS AND TRADES, Peter Stockham (ed.). Extremely rare 1807 volume describes to youngsters the crafts and trades of the day: brickmaker, weaver, dressmaker, bookbinder, ropemaker, saddler, many more. Quaint prose, charming illustrations for each craft. 20 black-and-white line illustrations. 192pp. 4⅝ x 6. 27293-1 Pa. $4.95

VICTORIAN FASHIONS AND COSTUMES FROM HARPER'S BAZAR, 1867–1898, Stella Blum (ed.). Day costumes, evening wear, sports clothes, shoes, hats, other accessories in over 1,000 detailed engravings. 320pp. 9⅜ x 12¼. 22990-4 Pa. $15.95

GUSTAV STICKLEY, THE CRAFTSMAN, Mary Ann Smith. Superb study surveys broad scope of Stickley's achievement, especially in architecture. Design philosophy, rise and fall of the Craftsman empire, descriptions and floor plans for many Craftsman houses, more. 86 black-and-white halftones. 31 line illustrations. Introduction 208pp. 6½ x 9¼. 27210-9 Pa. $9.95

THE LONG ISLAND RAIL ROAD IN EARLY PHOTOGRAPHS, Ron Ziel. Over 220 rare photos, informative text document origin (1844) and development of rail service on Long Island. Vintage views of early trains, locomotives, stations, passengers, crews, much more. Captions. 8⅞ x 11¾. 26301-0 Pa. $13.95

THE BOOK OF OLD SHIPS: From Egyptian Galleys to Clipper Ships, Henry B. Culver. Superb, authoritative history of sailing vessels, with 80 magnificent line illustrations. Galley, bark, caravel, longship, whaler, many more. Detailed, informative text on each vessel by noted naval historian. Introduction. 256pp. 5⅜ x 8½. 27332-6 Pa. $7.95

TEN BOOKS ON ARCHITECTURE, Vitruvius. The most important book ever written on architecture. Early Roman aesthetics, technology, classical orders, site selection, all other aspects. Morgan translation. 331pp. 5⅜ x 8½. 20645-9 Pa. $8.95

THE HUMAN FIGURE IN MOTION, Eadweard Muybridge. More than 4,500 stopped-action photos, in action series, showing undraped men, women, children jumping, lying down, throwing, sitting, wrestling, carrying, etc. 390pp. 7⅞ x 10⅝. 20204-6 Clothbd. $27.95

TREES OF THE EASTERN AND CENTRAL UNITED STATES AND CANADA, William M. Harlow. Best one-volume guide to 140 trees. Full descriptions, woodlore, range, etc. Over 600 illustrations. Handy size. 288pp. 4½ x 6⅜. 20395-6 Pa. $6.95

SONGS OF WESTERN BIRDS, Dr. Donald J. Borror. Complete song and call repertoire of 60 western species, including flycatchers, juncoes, cactus wrens, many more–includes fully illustrated booklet. Cassette and manual 99913-0 $8.95

GROWING AND USING HERBS AND SPICES, Milo Miloradovich. Versatile handbook provides all the information needed for cultivation and use of all the herbs and spices available in North America. 4 illustrations. Index. Glossary. 236pp. 5⅜ x 8½. 25058-X Pa. $6.95

BIG BOOK OF MAZES AND LABYRINTHS, Walter Shepherd. 50 mazes and labyrinths in all–classical, solid, ripple, and more–in one great volume. Perfect inexpensive puzzler for clever youngsters. Full solutions. 112pp. 8⅛ x 11. 22951-3 Pa. $4.95

PIANO TUNING, J. Cree Fischer. Clearest, best book for beginner, amateur. Simple repairs, raising dropped notes, tuning by easy method of flattened fifths. No previous skills needed. 4 illustrations. 201pp. 5⅜ x 8½. 23267-0 Pa. $6.95

A SOURCE BOOK IN THEATRICAL HISTORY, A. M. Nagler. Contemporary observers on acting, directing, make-up, costuming, stage props, machinery, scene design, from Ancient Greece to Chekhov. 611pp. 5⅜ x 8½. 20515-0 Pa. $12.95

THE COMPLETE NONSENSE OF EDWARD LEAR, Edward Lear. All nonsense limericks, zany alphabets, Owl and Pussycat, songs, nonsense botany, etc., illustrated by Lear. Total of 320pp. 5⅜ x 8½. (USO) 20167-8 Pa. $7.95

VICTORIAN PARLOUR POETRY: An Annotated Anthology, Michael R. Turner. 117 gems by Longfellow, Tennyson, Browning, many lesser-known poets. "The Village Blacksmith," "Curfew Must Not Ring Tonight," "Only a Baby Small," dozens more, often difficult to find elsewhere. Index of poets, titles, first lines. xxiii + 325pp. 5⅝ x 8¼. 27044-0 Pa. $8.95

DUBLINERS, James Joyce. Fifteen stories offer vivid, tightly focused observations of the lives of Dublin's poorer classes. At least one, "The Dead," is considered a masterpiece. Reprinted complete and unabridged from standard edition. 160pp. 5³⁄₁₆ x 8¼. 26870-5 Pa. $1.00

THE HAUNTED MONASTERY and THE CHINESE MAZE MURDERS, Robert van Gulik. Two full novels by van Gulik, set in 7th-century China, continue adventures of Judge Dee and his companions. An evil Taoist monastery, seemingly supernatural events; overgrown topiary maze hides strange crimes. 27 illustrations. 328pp. 5⅜ x 8½. 23502-5 Pa. $8.95

THE BOOK OF THE SACRED MAGIC OF ABRAMELIN THE MAGE, translated by S. MacGregor Mathers. Medieval manuscript of ceremonial magic. Basic document in Aleister Crowley, Golden Dawn groups. 268pp. 5⅜ x 8½. 23211-5 Pa. $9.95

NEW RUSSIAN-ENGLISH AND ENGLISH-RUSSIAN DICTIONARY, M. A. O'Brien. This is a remarkably handy Russian dictionary, containing a surprising amount of information, including over 70,000 entries. 366pp. 4½ x 6⅛. 20208-9 Pa. $9.95

HISTORIC HOMES OF THE AMERICAN PRESIDENTS, Second, Revised Edition, Irvin Haas. A traveler's guide to American Presidential homes, most open to the public, depicting and describing homes occupied by every American President from George Washington to George Bush. With visiting hours, admission charges, travel routes. 175 photographs. Index. 160pp. 8¼ x 11. 26751-2 Pa. $11.95

NEW YORK IN THE FORTIES, Andreas Feininger. 162 brilliant photographs by the well-known photographer, formerly with *Life* magazine. Commuters, shoppers, Times Square at night, much else from city at its peak. Captions by John von Hartz. 181pp. 9¼ x 10¾. 23585-8 Pa. $12.95

INDIAN SIGN LANGUAGE, William Tomkins. Over 525 signs developed by Sioux and other tribes. Written instructions and diagrams. Also 290 pictographs. 111pp. 6⅛ x 9¼. 22029-X Pa. $3.95

CATALOG OF DOVER BOOKS

ANATOMY: A Complete Guide for Artists, Joseph Sheppard. A master of figure drawing shows artists how to render human anatomy convincingly. Over 460 illustrations. 224pp. 8⅜ x 11¼. 27279-6 Pa. $11.95

MEDIEVAL CALLIGRAPHY: Its History and Technique, Marc Drogin. Spirited history, comprehensive instruction manual covers 13 styles (ca. 4th century thru 15th). Excellent photographs; directions for duplicating medieval techniques with modern tools. 224pp. 8⅜ x 11¼. 26142-5 Pa. $12.95

DRIED FLOWERS: How to Prepare Them, Sarah Whitlock and Martha Rankin. Complete instructions on how to use silica gel, meal and borax, perlite aggregate, sand and borax, glycerine and water to create attractive permanent flower arrangements. 12 illustrations. 32pp. 5⅜ x 8½. 21802-3 Pa. $1.00

EASY-TO-MAKE BIRD FEEDERS FOR WOODWORKERS, Scott D. Campbell. Detailed, simple-to-use guide for designing, constructing, caring for and using feeders. Text, illustrations for 12 classic and contemporary designs. 96pp. 5⅜ x 8½. 25847-5 Pa. $2.95

SCOTTISH WONDER TALES FROM MYTH AND LEGEND, Donald A. Mackenzie. 16 lively tales tell of giants rumbling down mountainsides, of a magic wand that turns stone pillars into warriors, of gods and goddesses, evil hags, powerful forces and more. 240pp. 5⅜ x 8½. 29677-6 Pa. $6.95

THE HISTORY OF UNDERCLOTHES, C. Willett Cunnington and Phyllis Cunnington. Fascinating, well-documented survey covering six centuries of English undergarments, enhanced with over 100 illustrations: 12th-century laced-up bodice, footed long drawers (1795), 19th-century bustles, 19th-century corsets for men, Victorian "bust improvers," much more. 272pp. 5⅝ x 8¼. 27124-2 Pa. $9.95

ARTS AND CRAFTS FURNITURE: The Complete Brooks Catalog of 1912, Brooks Manufacturing Co. Photos and detailed descriptions of more than 150 now very collectible furniture designs from the Arts and Crafts movement depict davenports, settees, buffets, desks, tables, chairs, bedsteads, dressers and more, all built of solid, quarter-sawed oak. Invaluable for students and enthusiasts of antiques, Americana and the decorative arts. 80pp. 6½ x 9¼. 27471-3 Pa. $8.95

HOW WE INVENTED THE AIRPLANE: An Illustrated History, Orville Wright. Fascinating firsthand account covers early experiments, construction of planes and motors, first flights, much more. Introduction and commentary by Fred C. Kelly. 76 photographs. 96pp. 8¼ x 11. 25662-6 Pa. $8.95

THE ARTS OF THE SAILOR: Knotting, Splicing and Ropework, Hervey Garrett Smith. Indispensable shipboard reference covers tools, basic knots and useful hitches; handsewing and canvas work, more. Over 100 illustrations. Delightful reading for sea lovers. 256pp. 5⅜ x 8½. 26440-8 Pa. $7.95

FRANK LLOYD WRIGHT'S FALLINGWATER: The House and Its History, Second, Revised Edition, Donald Hoffmann. A total revision–both in text and illustrations–of the standard document on Fallingwater, the boldest, most personal architectural statement of Wright's mature years, updated with valuable new material from the recently opened Frank Lloyd Wright Archives. "Fascinating"–*The New York Times*. 116 illustrations. 128pp. 9¼ x 10¾. 27430-6 Pa. $11.95

PHOTOGRAPHIC SKETCHBOOK OF THE CIVIL WAR, Alexander Gardner. 100 photos taken on field during the Civil War. Famous shots of Manassas Harper's Ferry, Lincoln, Richmond, slave pens, etc. 244pp. 10⅝ x 8¼. 22731-6 Pa. $9.95

FIVE ACRES AND INDEPENDENCE, Maurice G. Kains. Great back-to-the-land classic explains basics of self-sufficient farming. The one book to get. 95 illustrations. 397pp. 5⅜ x 8½. 20974-1 Pa. $7.95

SONGS OF EASTERN BIRDS, Dr. Donald J. Borror. Songs and calls of 60 species most common to eastern U.S.: warblers, woodpeckers, flycatchers, thrushes, larks, many more in high-quality recording. Cassette and manual 99912-2 $9.95

A MODERN HERBAL, Margaret Grieve. Much the fullest, most exact, most useful compilation of herbal material. Gigantic alphabetical encyclopedia, from aconite to zedoary, gives botanical information, medical properties, folklore, economic uses, much else. Indispensable to serious reader. 161 illustrations. 888pp. 6½ x 9¼. 2-vol. set. (USO)
Vol. I: 22798-7 Pa. $9.95
Vol. II: 22799-5 Pa. $9.95

HIDDEN TREASURE MAZE BOOK, Dave Phillips. Solve 34 challenging mazes accompanied by heroic tales of adventure. Evil dragons, people-eating plants, bloodthirsty giants, many more dangerous adversaries lurk at every twist and turn. 34 mazes, stories, solutions. 48pp. 8¼ x 11. 24566-7 Pa. $2.95

LETTERS OF W. A. MOZART, Wolfgang A. Mozart. Remarkable letters show bawdy wit, humor, imagination, musical insights, contemporary musical world; includes some letters from Leopold Mozart. 276pp. 5⅜ x 8½. 22859-2 Pa. $7.95

BASIC PRINCIPLES OF CLASSICAL BALLET, Agrippina Vaganova. Great Russian theoretician, teacher explains methods for teaching classical ballet. 118 illustrations. 175pp. 5⅜ x 8½. 22036-2 Pa. $5.95

THE JUMPING FROG, Mark Twain. Revenge edition. The original story of The Celebrated Jumping Frog of Calaveras County, a hapless French translation, and Twain's hilarious "retranslation" from the French. 12 illustrations. 66pp. 5⅜ x 8½. 22686-7 Pa. $3.95

BEST REMEMBERED POEMS, Martin Gardner (ed.). The 126 poems in this superb collection of 19th- and 20th-century British and American verse range from Shelley's "To a Skylark" to the impassioned "Renascence" of Edna St. Vincent Millay and to Edward Lear's whimsical "The Owl and the Pussycat." 224pp. 5⅜ x 8½. 27165-X Pa. $5.95

COMPLETE SONNETS, William Shakespeare. Over 150 exquisite poems deal with love, friendship, the tyranny of time, beauty's evanescence, death and other themes in language of remarkable power, precision and beauty. Glossary of archaic terms. 80pp. 5$\frac{3}{16}$ x 8¼. 26686-9 Pa. $1.00

BODIES IN A BOOKSHOP, R. T. Campbell. Challenging mystery of blackmail and murder with ingenious plot and superbly drawn characters. In the best tradition of British suspense fiction. 192pp. 5⅜ x 8½. 24720-1 Pa. $6.95

THE INFLUENCE OF SEA POWER UPON HISTORY, 1660–1783, A. T. Mahan. Influential classic of naval history and tactics still used as text in war colleges. First paperback edition. 4 maps. 24 battle plans. 640pp. 5⅜ x 8½. 25509-3 Pa. $14.95

THE STORY OF THE TITANIC AS TOLD BY ITS SURVIVORS, Jack Winocour (ed.). What it was really like. Panic, despair, shocking inefficiency, and a little heroism. More thrilling than any fictional account. 26 illustrations. 320pp. 5⅜ x 8½. 20610-6 Pa. $8.95

FAIRY AND FOLK TALES OF THE IRISH PEASANTRY, William Butler Yeats (ed.). Treasury of 64 tales from the twilight world of Celtic myth and legend: "The Soul Cages," "The Kildare Pooka," "King O'Toole and his Goose," many more. Introduction and Notes by W. B. Yeats. 352pp. 5⅜ x 8½. 26941-8 Pa. $8.95

BUDDHIST MAHAYANA TEXTS, E. B. Cowell and Others (eds.). Superb, accurate translations of basic documents in Mahayana Buddhism, highly important in history of religions. The Buddha-karita of Asvaghosha, Larger Sukhavativyuha, more. 448pp. 5⅜ x 8½. 25552-2 Pa. $12.95

ONE TWO THREE . . . INFINITY: Facts and Speculations of Science, George Gamow. Great physicist's fascinating, readable overview of contemporary science: number theory, relativity, fourth dimension, entropy, genes, atomic structure, much more. 128 illustrations. Index. 352pp. 5⅜ x 8½. 25664-2 Pa. $8.95

ENGINEERING IN HISTORY, Richard Shelton Kirby, et al. Broad, nontechnical survey of history's major technological advances: birth of Greek science, industrial revolution, electricity and applied science, 20th-century automation, much more. 181 illustrations. ". . . excellent . . ."–*Isis*. Bibliography. vii + 530pp. 5⅜ x 8¼. 26412-2 Pa. $14.95

DALÍ ON MODERN ART: The Cuckolds of Antiquated Modern Art, Salvador Dalí. Influential painter skewers modern art and its practitioners. Outrageous evaluations of Picasso, Cézanne, Turner, more. 15 renderings of paintings discussed. 44 calligraphic decorations by Dalí. 96pp. 5⅜ x 8½. (USO) 29220-7 Pa. $4.95

ANTIQUE PLAYING CARDS: A Pictorial History, Henry René D'Allemagne. Over 900 elaborate, decorative images from rare playing cards (14th–20th centuries): Bacchus, death, dancing dogs, hunting scenes, royal coats of arms, players cheating, much more. 96pp. 9¼ x 12¼. 29265-7 Pa. $12.95

MAKING FURNITURE MASTERPIECES: 30 Projects with Measured Drawings, Franklin H. Gottshall. Step-by-step instructions, illustrations for constructing handsome, useful pieces, among them a Sheraton desk, Chippendale chair, Spanish desk, Queen Anne table and a William and Mary dressing mirror. 224pp. 8⅛ x 11¼. 29338-6 Pa. $13.95

THE FOSSIL BOOK: A Record of Prehistoric Life, Patricia V. Rich et al. Profusely illustrated definitive guide covers everything from single-celled organisms and dinosaurs to birds and mammals and the interplay between climate and man. Over 1,500 illustrations. 760pp. 7½ x 10⅛. 29371-8 Pa. $29.95

Prices subject to change without notice.

Available at your book dealer or write for free catalog to Dept. GI, Dover Publications, Inc., 31 East 2nd St., Mineola, N.Y. 11501. Dover publishes more than 500 books each year on science, elementary and advanced mathematics, biology, music, art, literary history, social sciences and other areas.